## Pictures

This **picture shows** (a desert).
This is a **picture of** (the Sahara).

This is a **photo, a sketch, an aerial photograph, a drawing, a cartoon, showing** (Edinburgh).
Something (the castle) **is located**

- **in the top left corner.**
- **in the right-hand corner.**
- **in the bottom left corner.**
- **in the right-hand corner.**
- **in the middle / in the centre.**
- **in the foreground.**
- **in the background.**

The **photo**

- was taken **in** (Malta).
- was taken **at** (the seaside).
- was taken **on** (an oil rig).

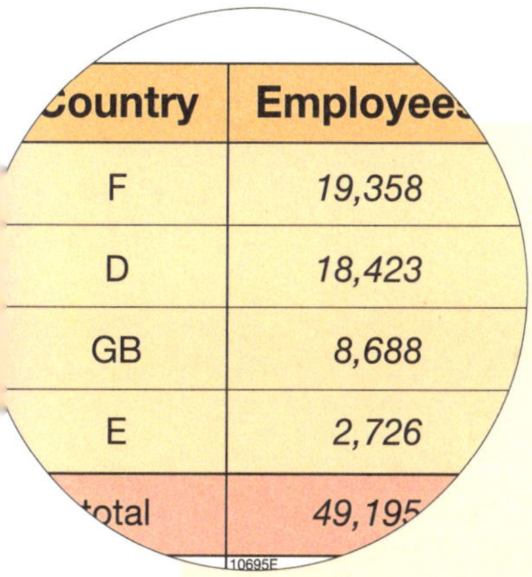

| ountry | Employee |
|---|---|
| F | 19,358 |
| D | 18,423 |
| GB | 8,688 |
| E | 2,726 |
| otal | 49,195 |

## Tables

This **table shows** (the number of workers).

The **figures in row** (two) **stand for** (the number of passengers).
The **figures in column** (A) **stand for** (the years).
The **figures in row** (3) **/ column** (2) **are positive / negative.**
The **number of** workers **has increased by** (10 %).
The **amount of** oil **has decreased by** (10 %).
The population **has doubled / trebled in** (3 years).

# DIERCKE GEOGRAPHY

## FOR BILINGUAL CLASSES

## BASIC

Moderation:
Prof. Dr. Reinhard Hoffmann

Autoren:
Matthew Appleby
Brigitte Dreymüller
Volker Friedrich
Dieter Haupt
Dirk Reischauer
Dimo M. Rischke
Horst Weible

westermann

Umschlag: London Tower Bridge



Druck: A⁴/ Jahr 2010
Alle Drucke der Serie A sind im Unterricht parallel verwendbar.

Verlagslektorat: Manfred Eiblmaier, Thomas Michael
Umschlaggestaltung und Layout: Heike Rieper, Thomas Schröder
Herstellung: Peter Kranz
Satz: PER Medien + Marketing GmbH, Braunschweig
Druck und Bindung: westermann druck GmbH, Braunschweig

ISBN 978-3-14-114010-1

# The World We Live in



**M1 The Earth and the Moon in space**

## Geography – What Is It About?

Geography brings you to the world around you. Geography describes the Earth.

M1 Glacier in Antarctica

M2 Rio de Janeiro

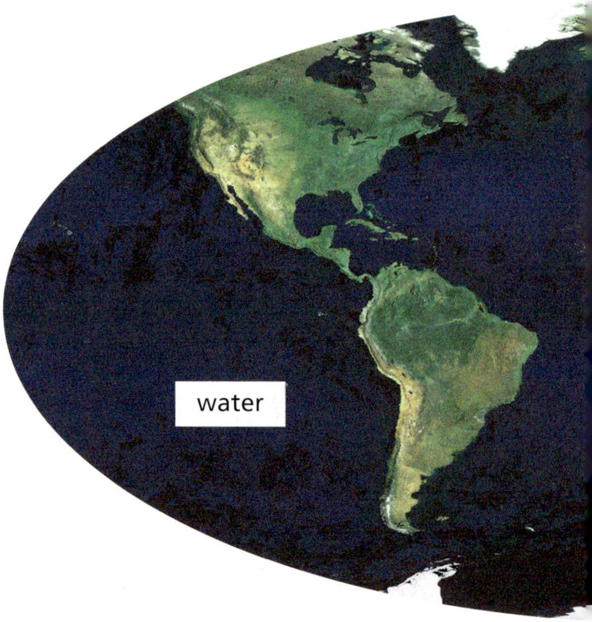

M3 Oxford Street, London

Geography comes from two Greek words:
*geos* = earth and
*graphein* = to describe

M4 In the Sahara Desert

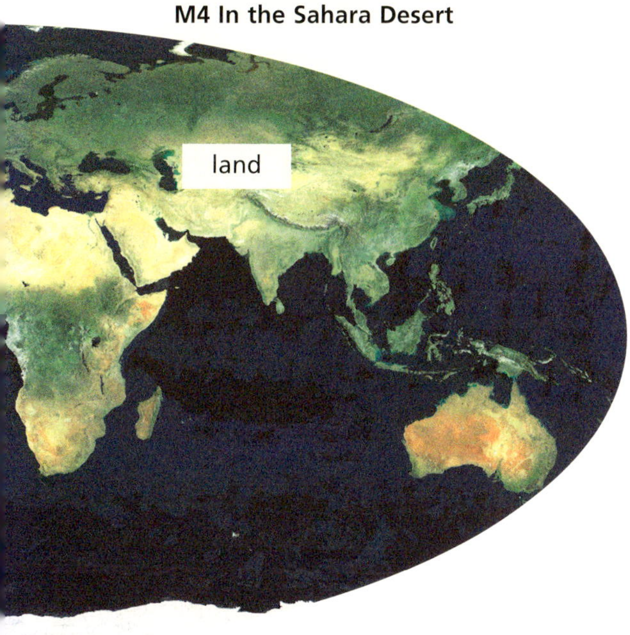

M5 Koala bear in a eucalyptus tree

M6 Rice field in Japan

**Tasks**

1 Describe your favourite picture.

## Geography in Our Classroom

In Geography we need many special tools. We work with different maps, textbooks, the Internet,...

### Tasks

**1** Make a list in your exercise book: **Basic Geography Tools**

wall map
encyclopaedia
ideo set
globe
the Internet
atlas

## The Earth – a Planet in Our Solar System

In the middle of our solar system is the Sun. Eight planets and some dwarf planets like Pluto circle the Sun. Our Earth is one of the planets. One trip around the Sun takes one year.
Moons circle most planets, the Earth has got only one moon.

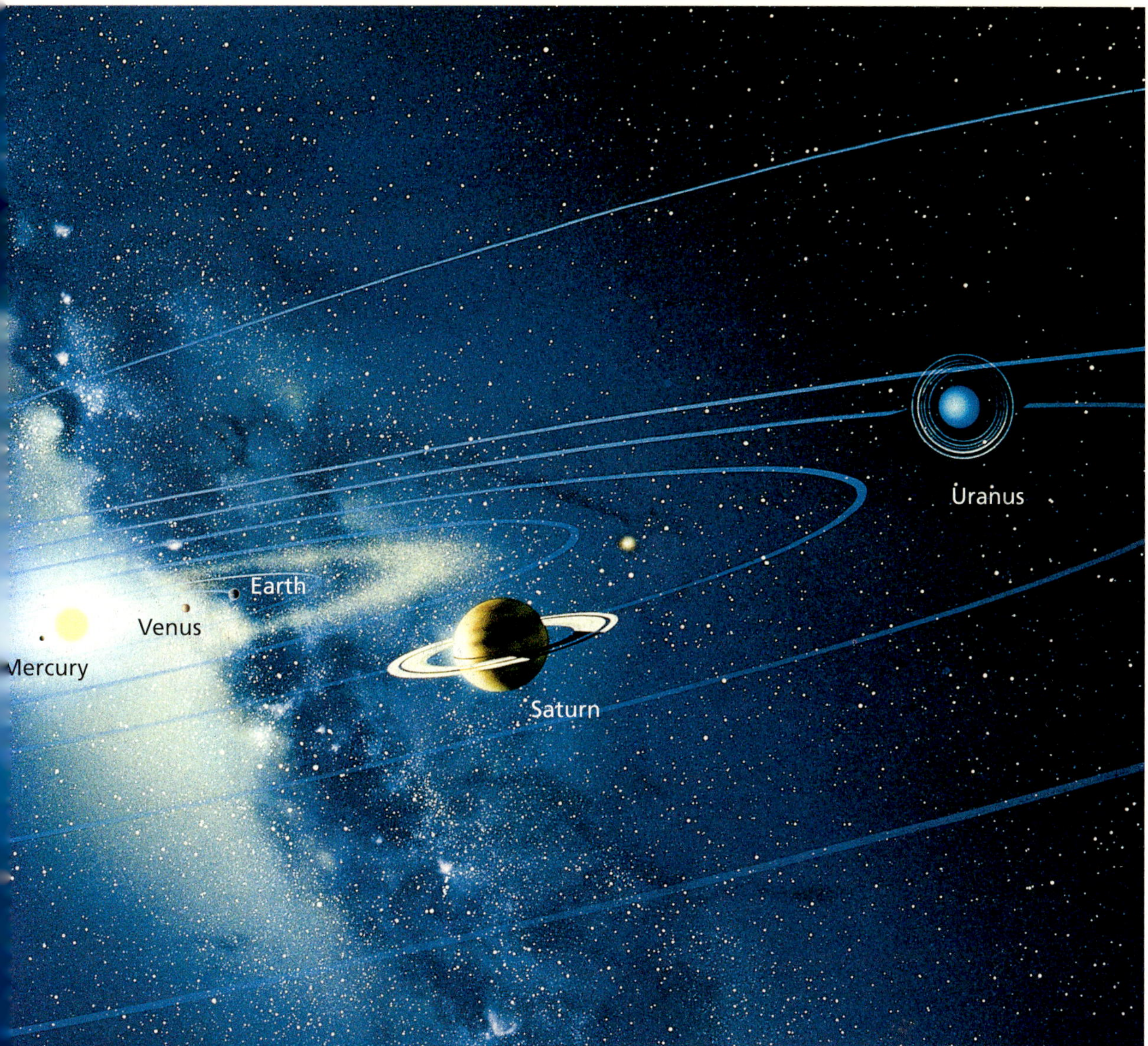

**Key terms:**
- solar system
- the Sun
- moon
- planet

## Tasks

**1** Use a basketball as a model of the Sun. What can we use as a model of the Earth: a tennis ball, a table tennis ball or just a pinhead?

Answer: It's a pinhead

## The Earth – the Planet Where We Live

Like the other planets the Earth spins on its own, invisible axis. The upper end of this axis is the North Pole. The bottom of the axis is the South Pole.
One spin round this axis takes 24 hours (M1) and causes night and day.

The axis has got a tilt, so the North Pole and the South Pole are not straight up and down (M2).

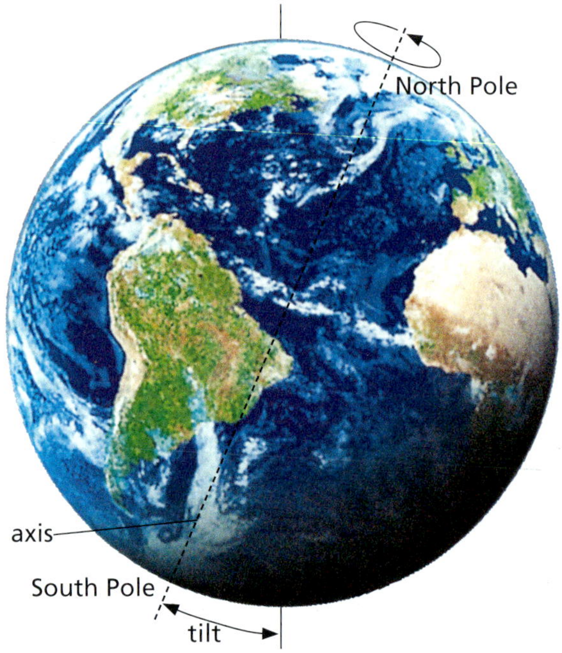


**M2 The Earth turns on an invisible axis.**

**M1 The spinning of the Earth**

A globe is a model of the Earth. We can see the continents, oceans, seas, and islands. With a globe you can show how the Earth spins on its axis.

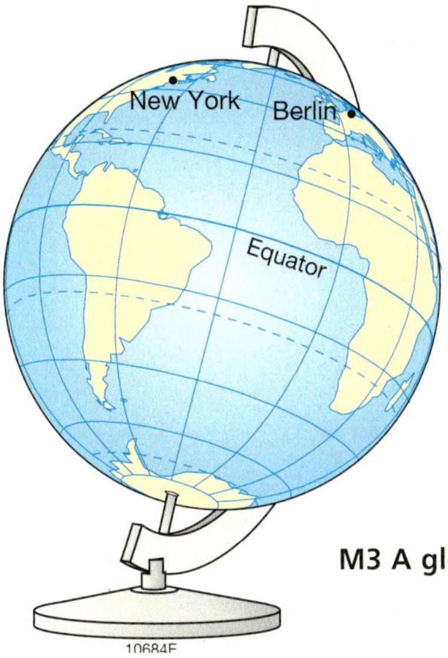


M3 A globe

## Tasks

**1** Put a globe on the teacher`s desk and turn the globe so that the class has got the same view as the astronaut in M1.

**2** Look at M1. Photo Number 9 is missing. What does it look like?

**Summary**

The Earth spins on an invisible axis.
One spin takes a day.
A globe is a model of the Earth.

**Key terms:**

- axis
- globe
- the North Pole
- the South Pole
- to spin
- tilt

# The Global Grid

A cold night in the spring of 1912.
The Titanic, the biggest ship of that time, has hit an iceberg in the North Atlantic and is sinking.
Many people are in the cold water. There are not enough life-boats.

**M1 The Titanic is sinking**

**M2 Where is the place of the accident? How can we locate the place?**

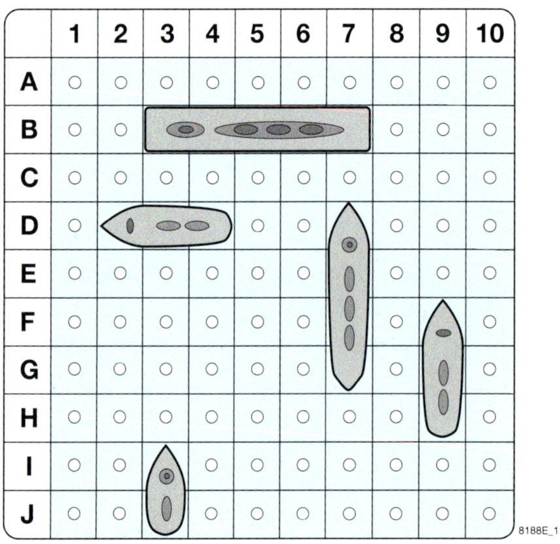


**M3 Battleships**

**M4 The Earth in a net**

# How to Locate Places

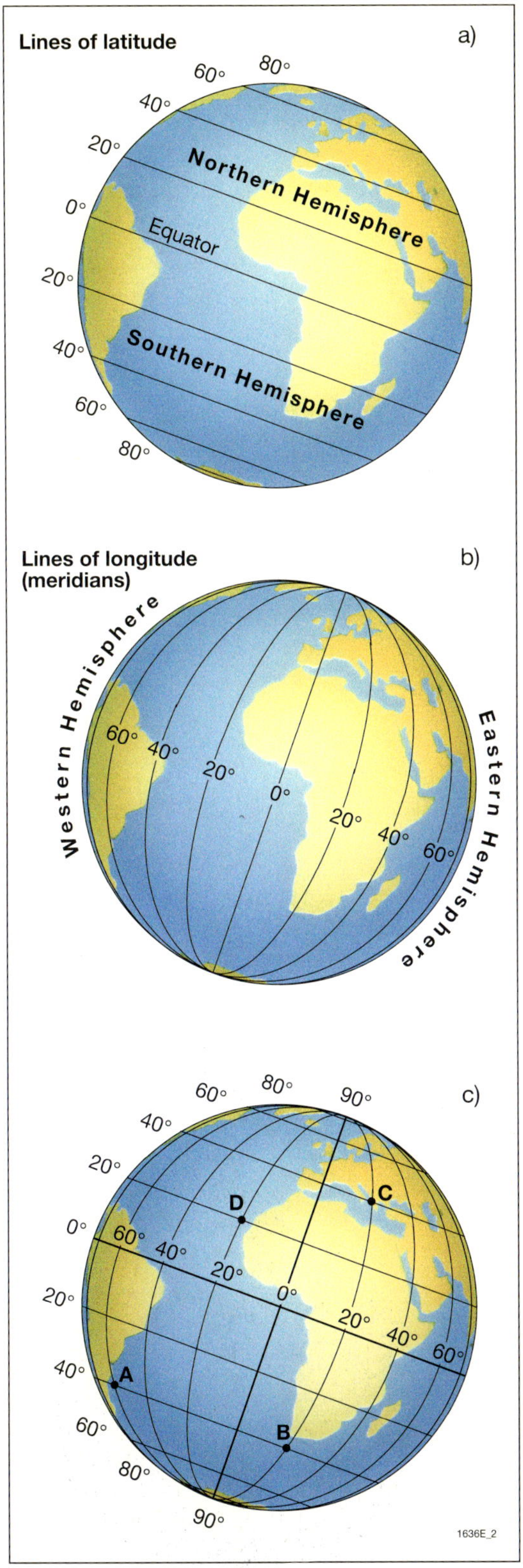


M5 The global grid

## How to Describe a Position on Earth

To locate places on Earth we use two sets of lines:

Lines of latitude
They run round the Earth from east to west. The Equator is at 0 degrees, the North Pole is at 90 degrees north, the South Pole is at 90 degrees south (M5).

Lines of longitude
They make a circle round the Earth and cross through the poles. Half of the circle from one pole to the other is a meridian. The Prime Meridian runs through Greenwich (London) at 0 degrees.

Together these two sets of lines form a grid. We call it the global grid, which helps us to describe where a certain point is.

The exact location of a place is where lines of longitude and latitude meet.

Example:
Where is place 'D' in M5c?
**Say:** Place 'D' is at twenty degrees north and twenty degrees west.
**Write:** Place 'D' is at 20° N / 20° W.

### 1 Tasks

Locate places A, B and C in M5.

**Summary**

We use lines of latitude and lines of longitude in the global grid to locate places.

**Key terms:**
- to describe a position
- Equator
- global grid
- lines of latitude
- lines of longitude
- to locate a position
- location
- Prime Meridian

## From the Globe to the Map

**Step 1:**
The globe is a model of the Earth, but geographers normally use maps. On a globe you cannot see all parts of the Earth at the same time. How can you transform the round surface of the globe into a flat map?

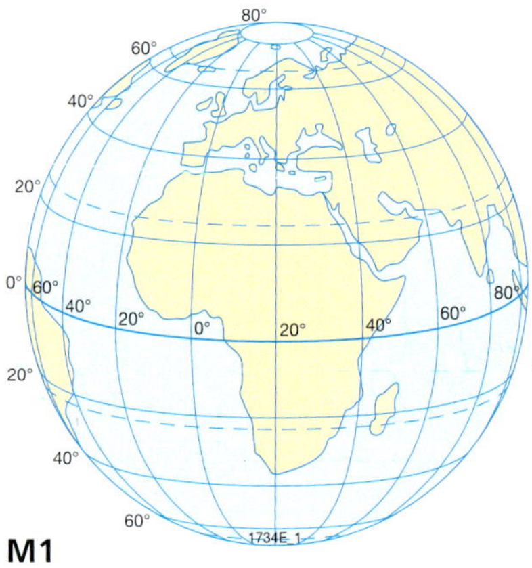


M1

**Step 2:**
Think of the globe as an orange and 'peel' it.

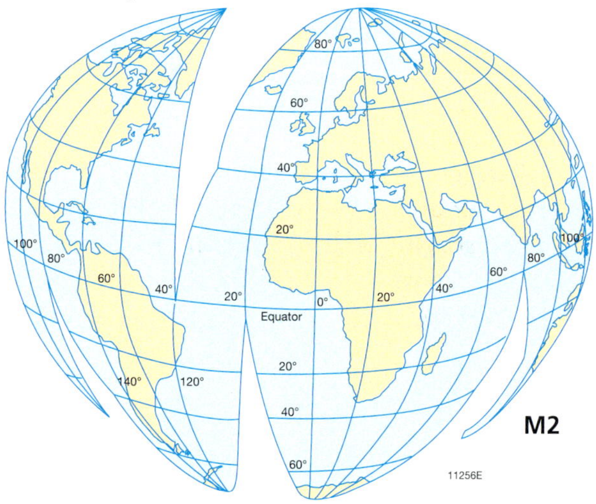


M2

**Step 3:**
When you flatten it, there are big gaps!

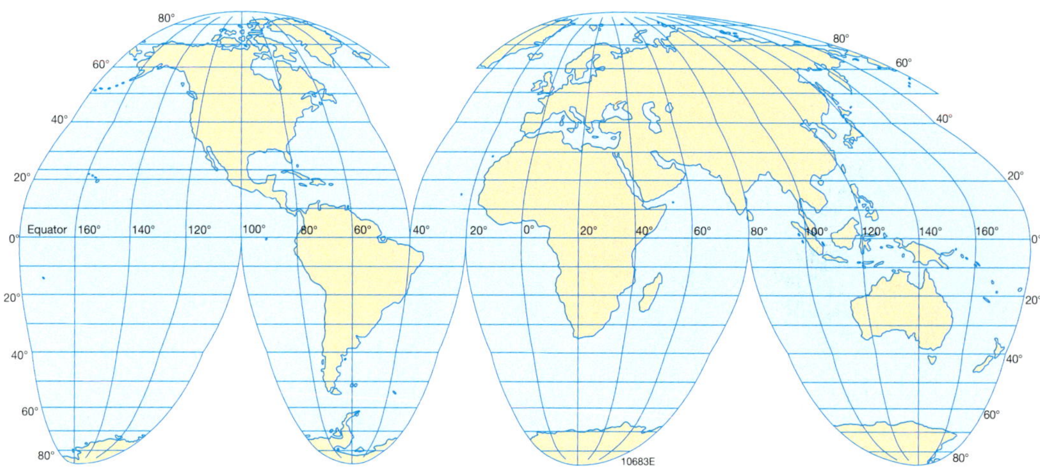


M3

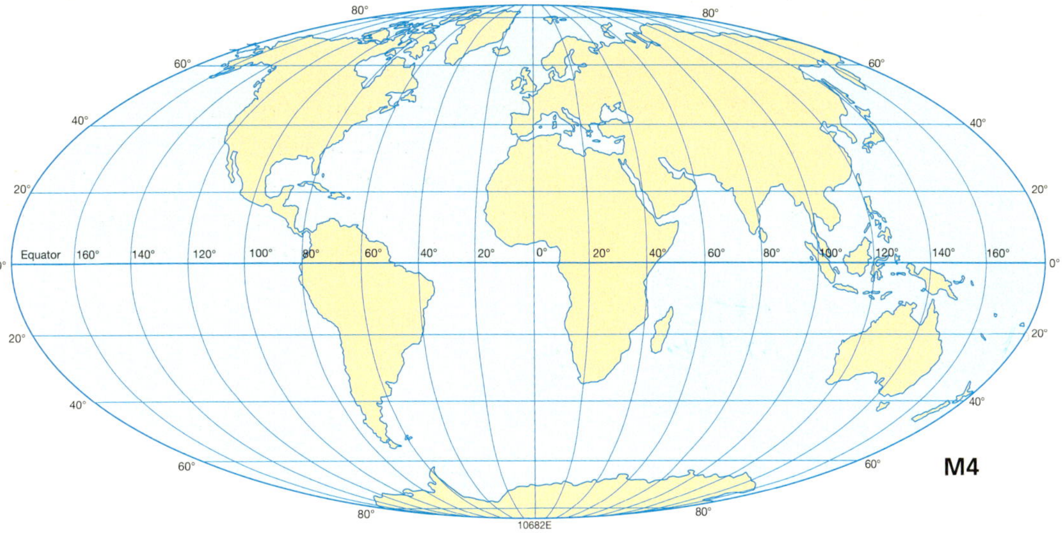


M4

**Step 4:**

Map-makers (cartographers) stretch parts of the Earth's surface to fill these gaps. So the size and the shape of oceans and continents are no longer correct. They can be different on different maps.

The world map (M4) has got a grid system, too, like the globe.
Normally, north is at the top of a map. With the help of a compass you can find north in the field.

## How to Work with a Compass

We use the compass to find directions. North, south, east and west are the points of the compass (M5).
The coloured end of the compass needle always points to the north. So, turn the compass until North (N) lines up with this end of the needle to find where north is.

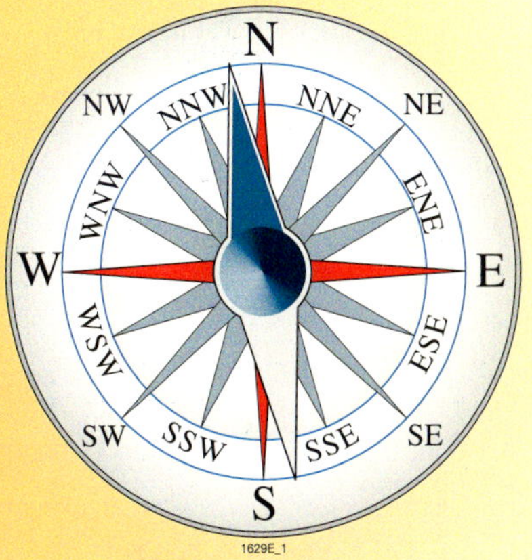


## 1 Tasks

Take a map of your area and put it on the floor. Make sure that the northern direction on the map lines up with the northern direction in reality. Use a compass.

**Key terms:**
- points of the compass (north, south, east, west)
- direction
- map
- shape
- size

**Summary**

Like a globe maps show the surface of the Earth. The size and shape of continents and oceans on maps are not correct. Normally, north is at the top of a map.

**M1 Physical map**

# How to Read a Physical Map

There are many different kinds of maps in an atlas. You may already know road maps or street maps of a town.

**M2 Boy reading a street map**

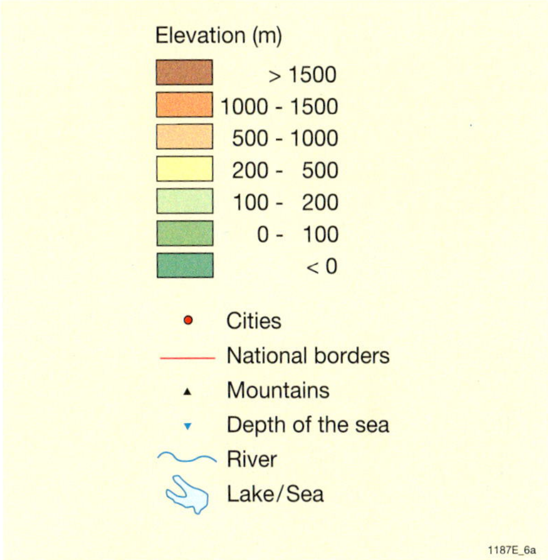


**M3 Map Key: Maps use symbols and colours for many physical facts**

A very important map is a physical map. It shows:
– the height (elevation) of the land;
– the location of cities, borders, mountains;
– the depth of the sea;
– the location of rivers and lakes.
You can also see lines of longitude and lines of latitude.

Every map has got a title. It often says which part of the world the map shows and what kind of map it is (M1).

On a physical map the colours green, yellow or brown show how high the land is. The colours white or blue stand for rivers, lakes, and seas. You find this information in the key (M3).

A map has got a scale. Map M1 has got a scale of 1 : 25,000,000, which means 1 cm on the map is 25 million centimetres (250 km) in reality.
A scale line makes it easier to measure distances.

## Tasks

**1** In M1, the distance from Madrid to Bern is 5 cm. What is the distance in kilometres?

**2** Take a ruler and measure the distance from
- Dublin to Stockholm
- London to Warsaw.
What is the distance in kilometres?

**Summary**

Maps have got a title, a scale or scale line, and a key. They help us to read a physical map.

**Key terms:**

- bar
- depth
- elevation/ height
- key
- physical map
- scale (line)

## Continents and Oceans

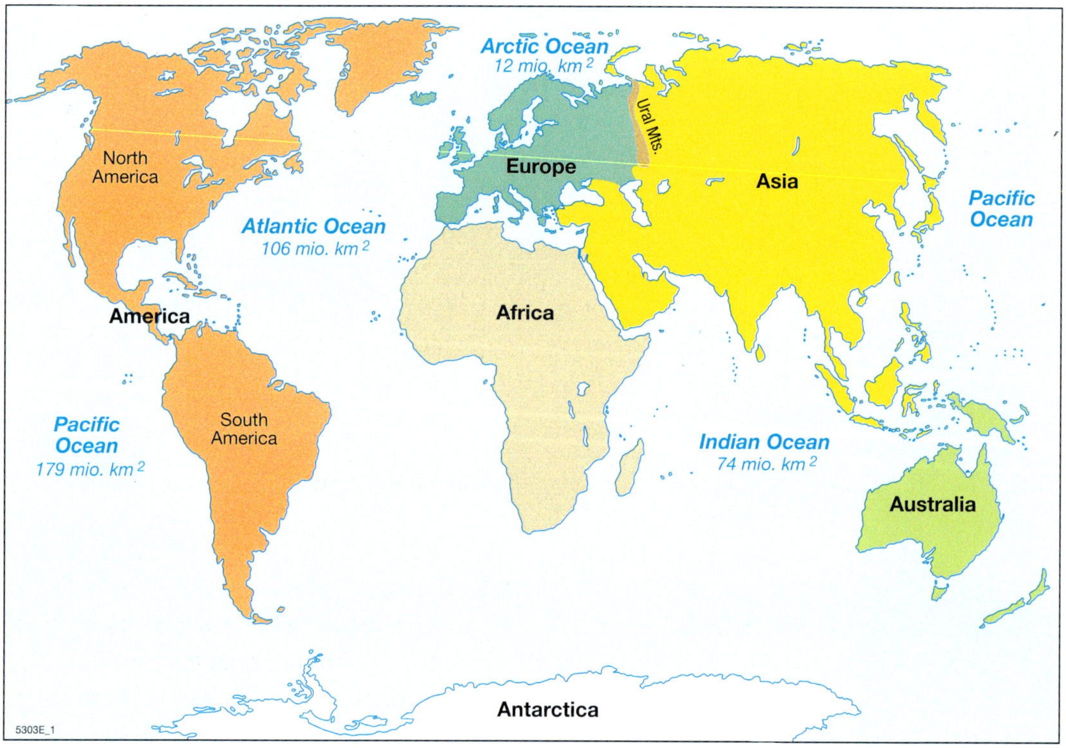


**M1 Continents and Oceans**

### The Earth – the 'Blue Planet'

We live on a planet of oceans. Water covers most of the Earth's surface.

Although we give different names to the four oceans – the Pacific, the Atlantic, the Indian, and the Arctic Ocean – in fact these four oceans are one big ocean.

The big landmasses are the continents. There are six: Africa, America, Antarctica, Asia, Australia, and Europe.

Most landmasses are north of the Equator, in the Northern Hemisphere.

**M2 Water covers most of the Earth's surface**

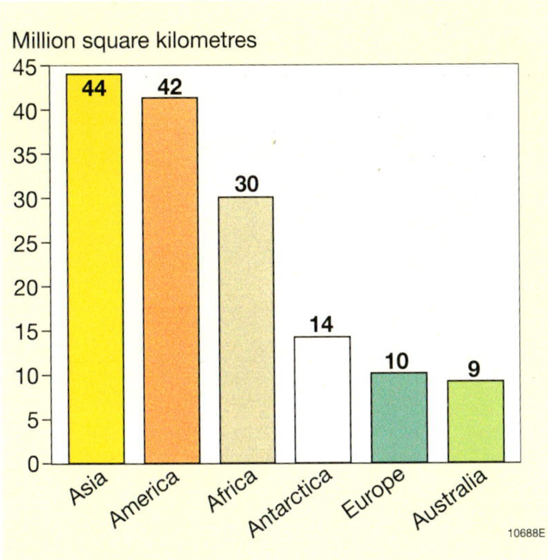


**M3 Bar graph: sizes of continents**

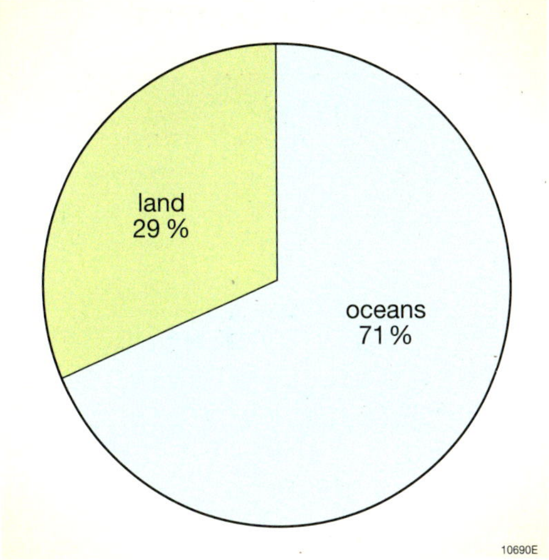


**M4 Pie chart: land and water on Earth**

## Tasks

**1** Use M1 and describe the location of continents and oceans.

**2** Explain the term 'the Blue Planet'.

**3** In M1 you find the sizes of the oceans. Take your exercise book and draw a bar graph.

**4** What is larger: the total landmass or the Pacific Ocean?

**Summary**

Most of the Earth's surface is water. The continents cover not much more than a quarter of the Earth's surface.

**Key terms:**

- bar graph
- continent
- to cover
- landmass
- ocean
- pie chart
- size
- surface

**M5 The big landmasses are continents**

## What Makes Day and Night?

**M1 Day and night from space – with Potsdamer Platz in Berlin in darkness and New York's Central Park still in broad daylight**

The Earth spins on its axis once every 24 hours from west to east. The Sun is the Earth's only source of light. This spinning causes day and night as the Sun only shines on one half of the Earth. The part of the Earth away from the Sun is in darkness, or has got nighttime, the other side has got daytime.

At noon, 12 o'clock, the Sun is at its highest point in the sky. Normally places that are directly north or south of each other have the same time.

### What's the time?

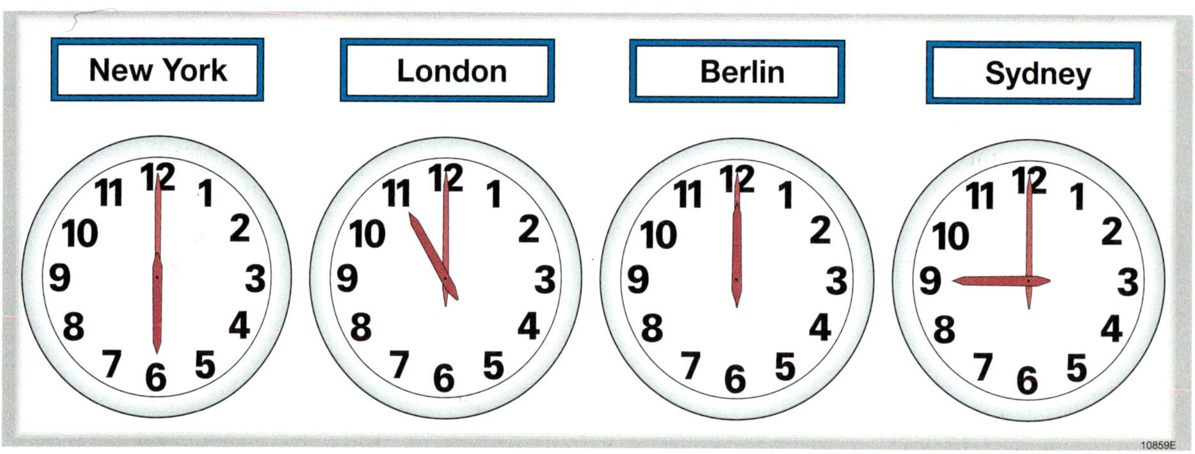


**M2 Time zones**

## What are Time Zones?

As some parts of the world are in light and other parts are in darkness, it can be early morning in some places and evening in others. For this reason there are time zones.
The time zones run nearly parallel to the lines of longitude. For every 15 degrees of longitude there is one hour time difference. The International Date Line nearly follows the 180° meridian. When you cross it, the date changes.

**M3 Video conference between the continents**

### Tasks

**1** Use your atlas to find out how many time zones there are in Russia, in Germany and in the USA. What is the maximum time difference?

**2** It is 6 p.m. in Amsterdam. Lisanne wants to start a video conference with Sandrito (M3). What must she think about?

**3** Is there a time which is ideal for all three people (M3)? Explain.

**Summary**

The spin of the earth on its axis causes day and night. So there are different time zones that run nearly parallel to the lines of longitude.

**Key terms:**
- daytime
- International Date Line
- nighttime
- to spin
- time zone

## Reasons for Seasons

M1 Spring

M2 Summer

M3 Autumn

M4 Winter

The four seasons – spring, summer, autumn, and winter – as we know them mean long and short days, falling leaves, etc. Why is that so?

**The Earth travels around the Sun**
On our planet, seasons are caused by the way the Earth takes around the Sun. As the Earth's axis is tilted, different parts of the Earth face the Sun in the course of the travel. This journey lasts one year.

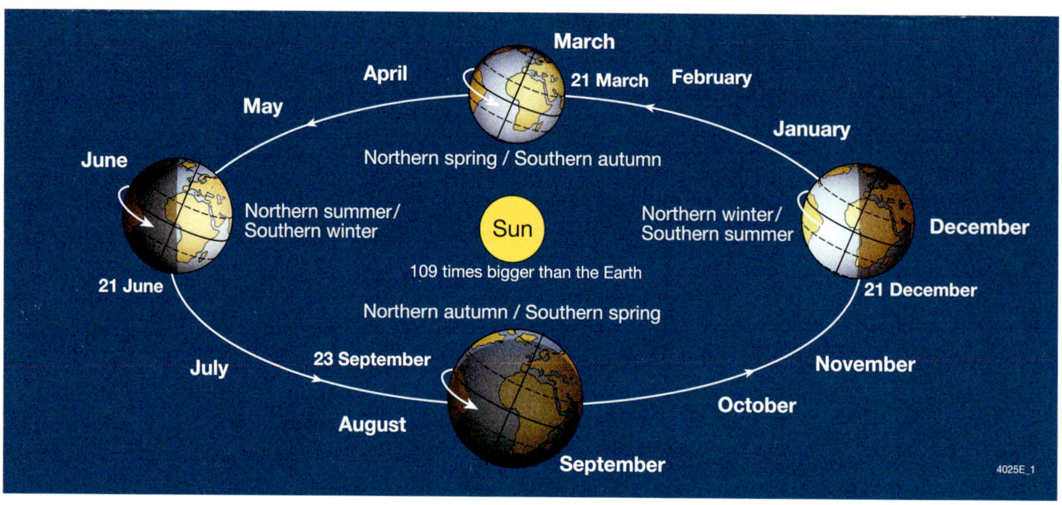


M5 The way of the Earth around the Sun

**The Earth is a Sphere**
The Sun's rays heat the Earth. Some places get more heat than other places. Vertical rays heat up the Earth's surface more than oblique rays, because they concentrate on a smaller area (M6). When vertical rays reach the Northern Hemisphere, it is warm there and we call this summer. When no vertical rays reach the Northern Hemisphere, it is cold there and we call it winter.

Some places get so little heat from the Sun that ice covers them the whole year. Some places get so much heat from the Sun that it is hot the whole year.

The seasons on the northern half of the earth and the southern half of the earth are different. When it is summer in the Southern Hemisphere, it is winter in the Northern Hemisphere. When it is winter in the Southern Hemisphere, it is summer in the Northern Hemisphere.

Between March and September it is the warm season for the northern half of our planet. Between September and March it is the warm season for the southern half of our planet.

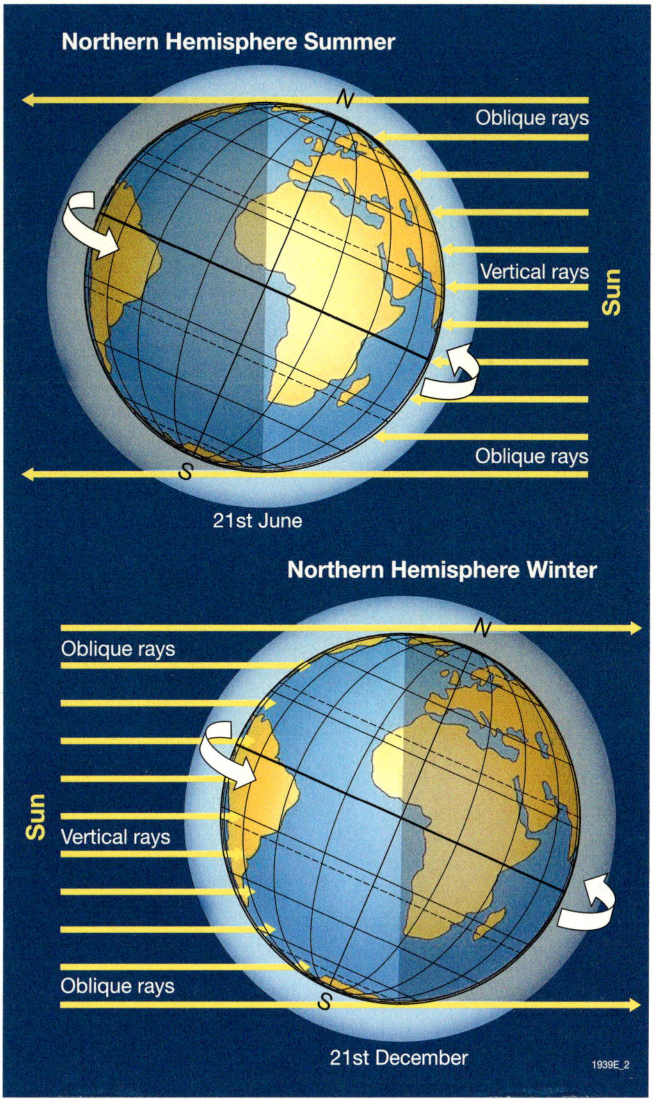


**M6 Summer and winter in the Northern Hemisphere**

**Summary**

The travel of the Earth around the Sun, its shape as a sphere and its tilted axis make the seasons: spring, summer, autumn, and winter.

## Tasks

**1** Describe how your daily life changes during the four seasons (M1–M4).

**2** Look at M6! Describe differences between the Northern Hemisphere in summer and in winter.

**Key terms:**
- Earth's axis
- Northern Hemisphere
- oblique
- season
- Southern Hemisphere
- tilt(ed)
- vertical

## What is Weather?

**M1 Satellite map of a hurricane**

**M2 The weather machine**

## What Does the Weather Map Show?

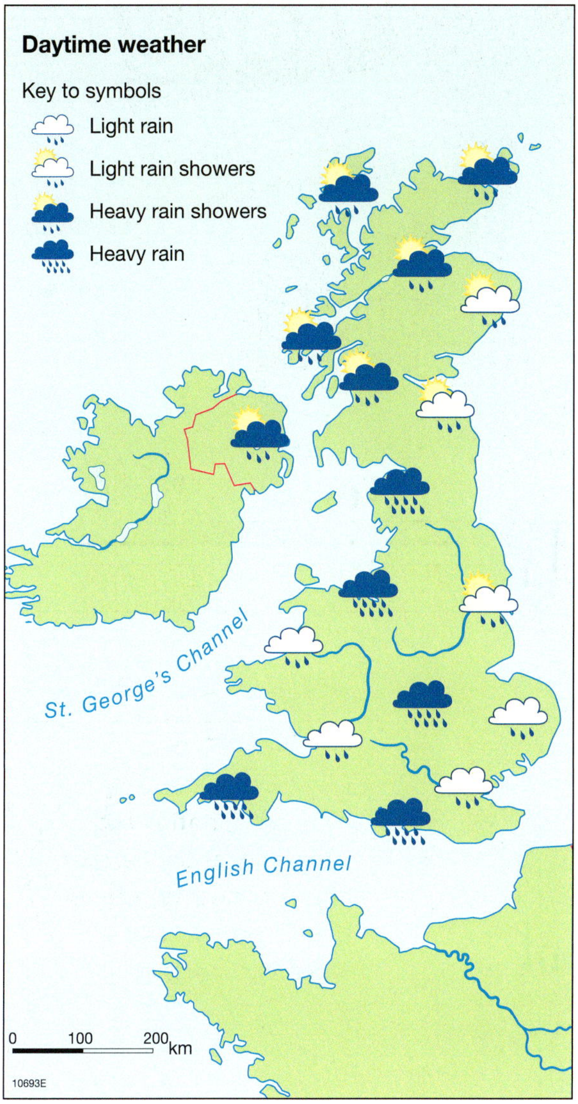


**M3 UK weather map with the weather situation on a day in January**

### Tasks

**1** Look at the weather map (M3) and describe the weather in the different parts of the UK (England, Wales, Scotland and Northern Ireland).

**2** What are the reasons why people want and need a good weather forecast?

**Summary**

To describe the weather you look at different weather elements, e.g.

- precipitation
- temperature
- wind
- clouds
- air pressure

You can use a weather map to describe and forecast the weather.

**Key terms:**
- forecast
- precipitation
- temperature
- weather

# How Can We Measure and Record … Temperature?

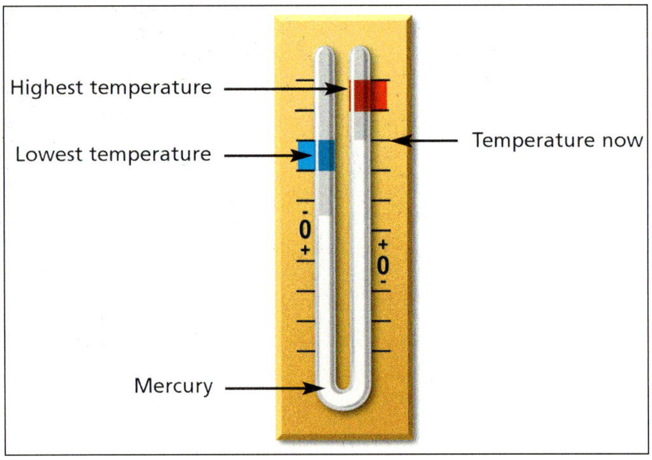


M1 Maximum and minimum thermometer

M2 Stevenson's Screen

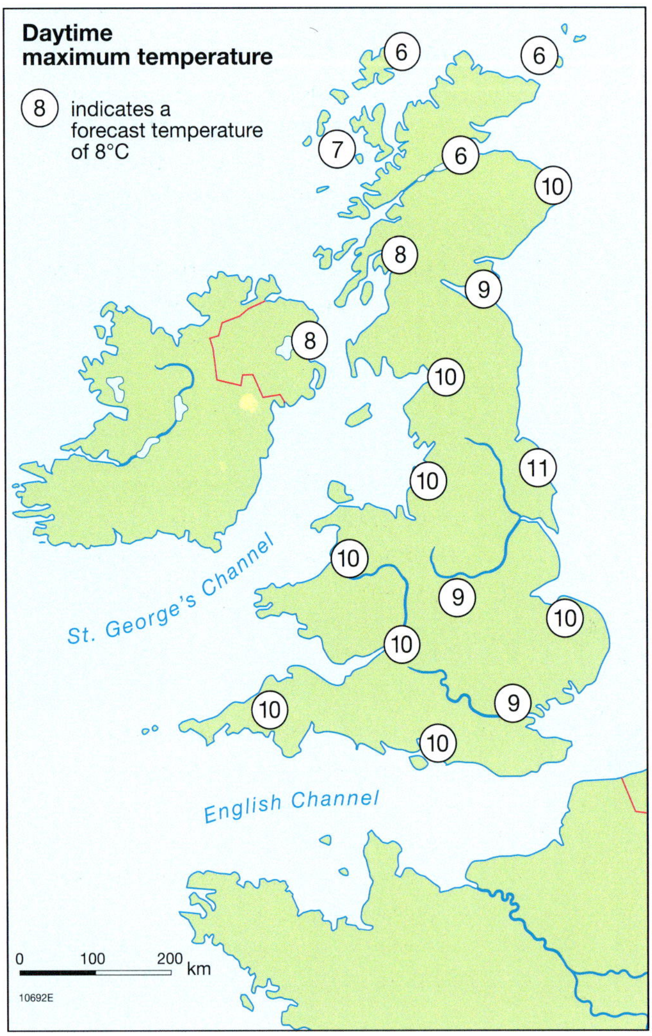


M3 TV-Map of daytime maximum temperatures

# How Can We Measure and Record ... Wind?

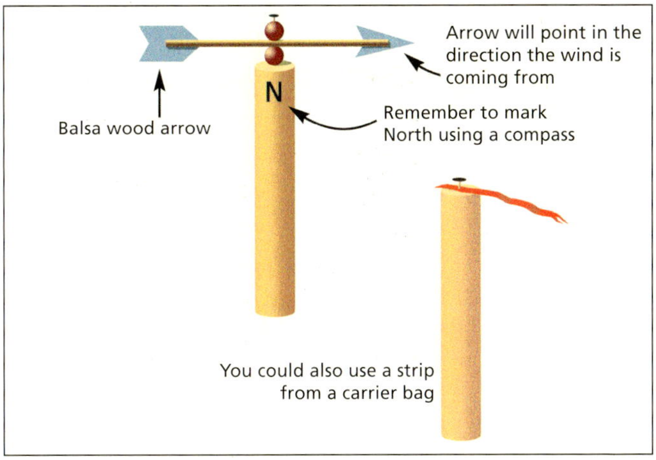


**M4 Self-made wind vane to measure wind direction**

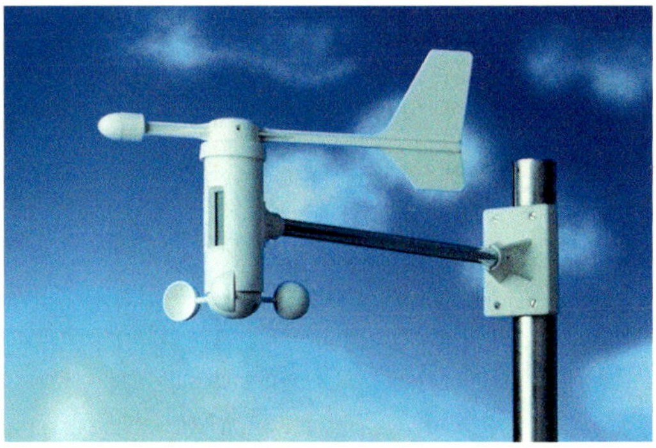

**M5 Cup anemometer to measure wind speed. The cups of the anemometer spin round. The number of revolutions in a given time gives the wind speed.**

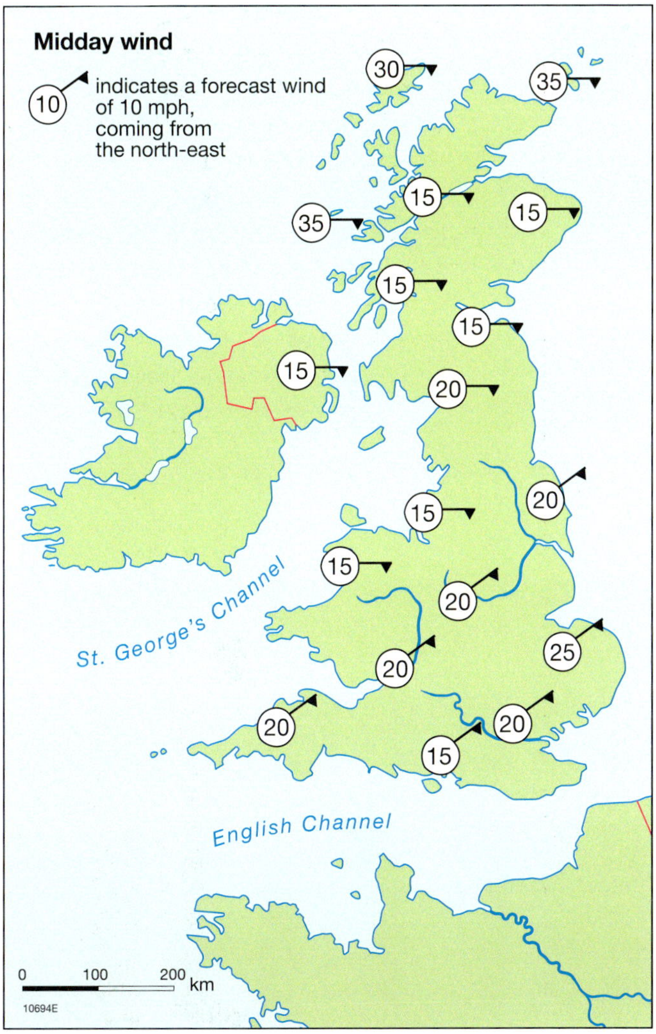


**M6 TV-Map of wind direction and speed**

## Tasks

**1** Use a thermometer to measure and record the temperature of the same hour of the day for one week. Make a table and describe what you find out.

**2** Compare the temperatures and the wind in the north of Scotland with those in Northern Ireland.

**Key terms:**

- anemometer
- maximum and minimum thermometer
- Stevenson's Screen
- wind speed
- wind vane

**Summary**

A minimum and maximum thermometer measures the highest and lowest temperatures of a period of time.
A wind vane helps you to find out the wind direction.
With an anemometer you can measure the wind speed.

# How Can We Measure and Record ... Precipitation?

Precipitation can look quite different. In the dry African savanna, it often does not rain for a long time, but especially in summer heavy showers can happen. However, in Germany the precipitation falls down in the form of snow if the temperatures are low enough in winter.

**M1 A heavy shower in the African savanna**

**M2 Snowfall in Europe**

## Types of precipitation

If the droplets in the clouds become too big precipitation starts.
**Rain** droplets vary in size from 0.5 mm to more than 5 mm in diameter.
**Drizzle** droplets are smaller than 0.5 mm and sometimes fall from fog.
**Fog** has very small droplets and is like a cloud that has touched the ground.
**Showers** are precipitation that suddenly start and end and quickly change in intensity.
**Snow** forms when water in the air turns to ice. Snow flakes are formed by very small water crystals.
**Hail** is formed when water droplets freeze in high, very cold clouds. Hailstones can also get bigger when they stick together. Usually they are only as big as a pea, but sometimes they become as big as tennis balls.

To measure precipitation you can buy a rain gauge or you can make one yourself with a plastic soda bottle.

**M3 Fog in America**

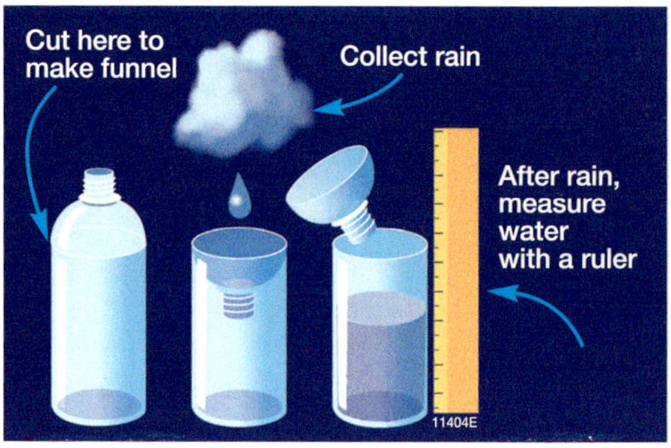



**M4 Self-made rain gauge**

## How Can We Measure and Record ... Clouds?

**M5 A nearly cloudless sky on a sunny day**

**M6 Dark clouds cover the sky completely. Such an overcast sky often means rain.**

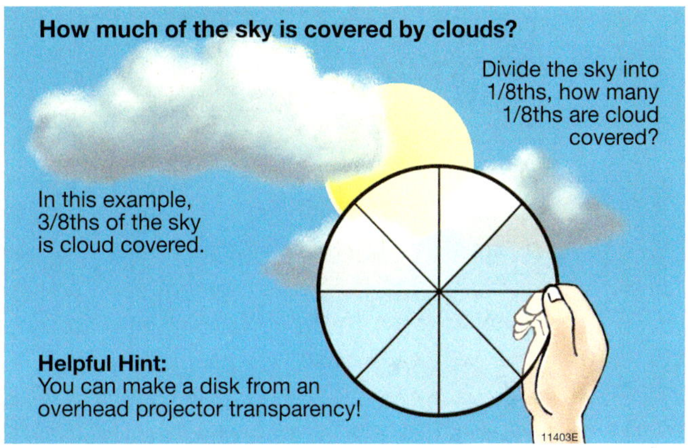


**M7 Cloud cover disk**

**Cloud cover:**
**cloudless** – of course, no clouds
**fine** – some clouds that cover less than half the sky
**half-covered** – of course, clouds cover half the sky
**cloudy** – clouds cover more than half the sky but not all
**overcast** – clouds cover the sky completely

### Tasks

**1** Make your own rain gauge or your own cloud cover disk and record the precipitation or the cloud cover for one week!

**Key terms:**
- cloud cover
- drizzle
- fog
- hail
- overcast
- rain
- rain gauge
- shower
- snow

**Summary**

You can characterize precipitation by talking about the different types of precipitation, e. g. rain, drizzle, snow, or by looking at the amount of precipitation.
To describe and record clouds you can measure how much of the sky is covered by clouds.

# What Do Landscapes Look Like?

M1 A plain

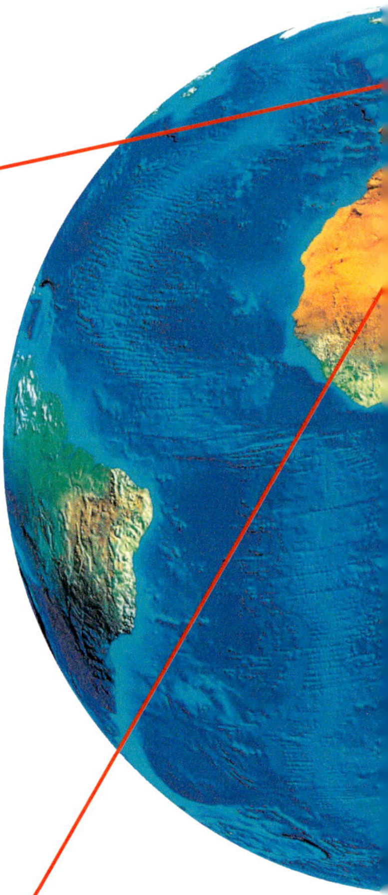

**On plains**
there is a flat surface with only a few small hills. In Central Europe you can find trees, long or short grasses. As you can easily use machines, the flat surface is ideal to grow crops.

**In mountains**
there are low temperatures throughout the year and most precipitation falls as snow. So in the highest regions there is no vegetation. And of course, as the temperatures are so low and the mountains are so steep you cannot grow crops in the higher parts.

M2 Mountains

## Tasks

**1** Compare the possibilities to grow crops in the landscapes you find on these two pages. Give reasons for the differences.

**Key terms:**
- dense
- desert
- forest
- grassland
- landscape
- plain
- rainforest
- relief
- sparse
- vegetatio

M3 A desert

**In deserts**
it is often very hot with only very little precipitation. Sometimes it does not even rain for years.
The vegetation is mostly rather sparse. Large areas are just covered by sand or stones. In the desert it is too dry to grow crops.

**In rainforests**
near the Equator it is warm and rains a lot throughout the year. There is a very dense forest vegetation with many different plants and animals. You can grow crops throughout the year.

M4 Tropical rainforest

**Summary**

As temperature, precipitation and relief can be very different from place to place, you can find different landcapes on Earth, too.
Some important landscapes are:
- plains
- mountains
- deserts
- rainforests

## 6,000,000,000 People

The Earth has got more than six billion inhabitants. They live permanently on all the continents except Antarctica. More than half of the people live in Asia. Australia has got the smallest population. Many live in large towns, but many also live in villages.

M1 Tourists in Benidorm, Spain

M2 Shopping in Greenland

M3 Computer class in Pennsylvania, USA

M4 Street children in Rio de Janeiro, Brazil

Many different peoples with different cultures live on Earth.

- More than one billion speak Mandarine, a Chinese language. Then comes English as number two.
- There are also many religions. Some examples are Christianity, Hinduism, Islam.

M5 Cooking in Malawi

M6 Fastfood in Peking, China

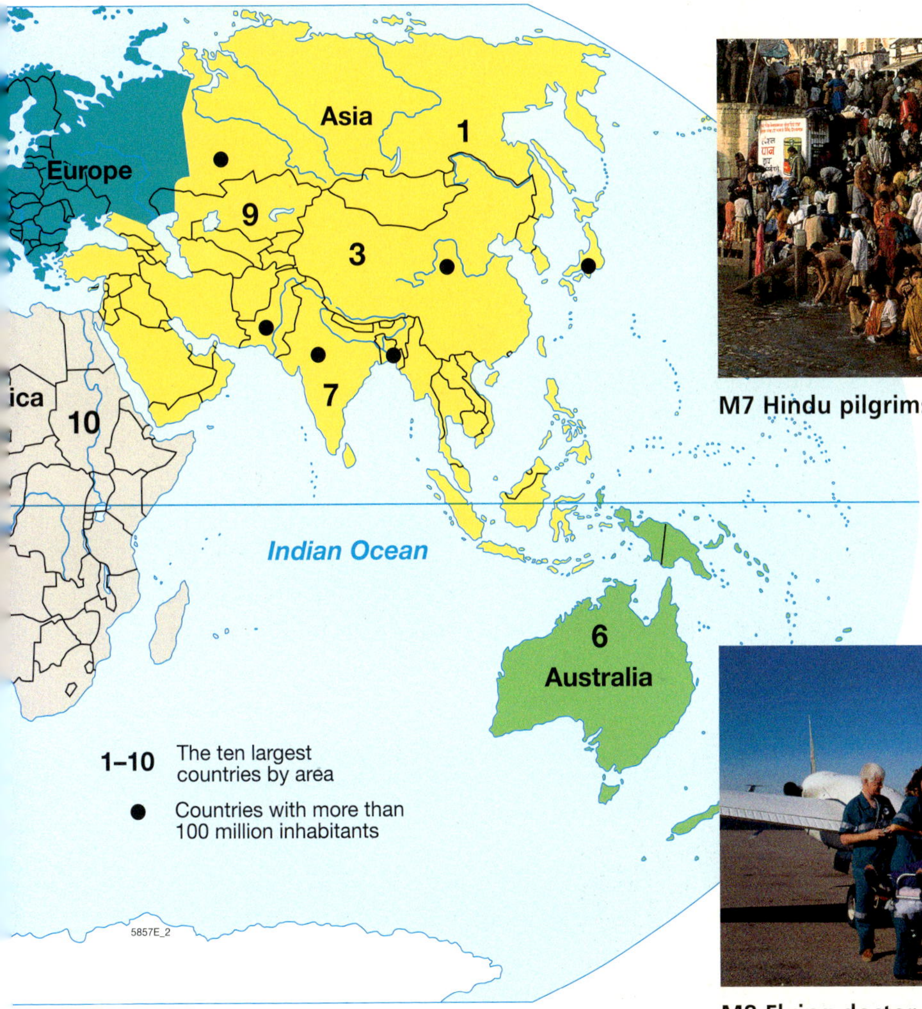

M7 Hindu pilgrims on the Ganges, India

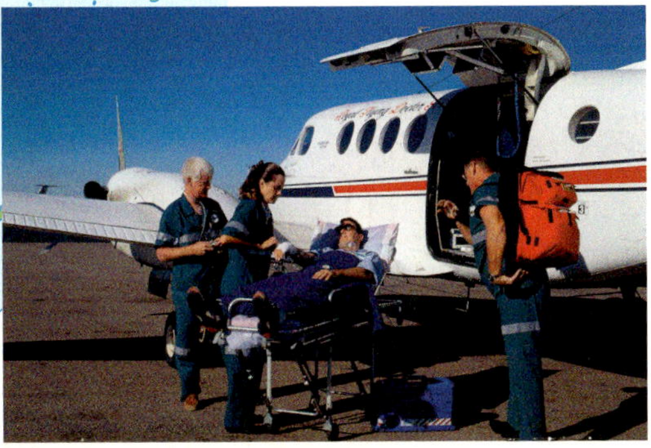

M8 Flying doctor, Australia

The people on Earth have not got equal living conditions.

- 25 per cent of the Earth's population has got no home. There are similar problems with water, food, schools, and medicine.
- Only one out of 12 persons has got an Internet connection at home.

**Key terms:**

– inhabitant
– language
– living conditions
– peoples
– population
– religion

# Distribution of Population

M1 shows two areas of one square kilometre each with 48 people. So the population density is 48 people per square kilometre.
In picture A these 48 persons do not live in groups, but all over the area. So area A has got an even distribution of population.
In area B the population shows an uneven distribution. There are two groups of persons in two parts. In other parts of B there are no people.

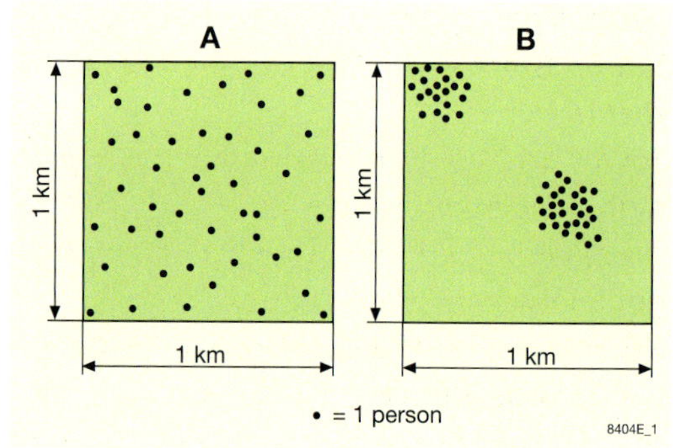


**M1 Examples of population distribution**

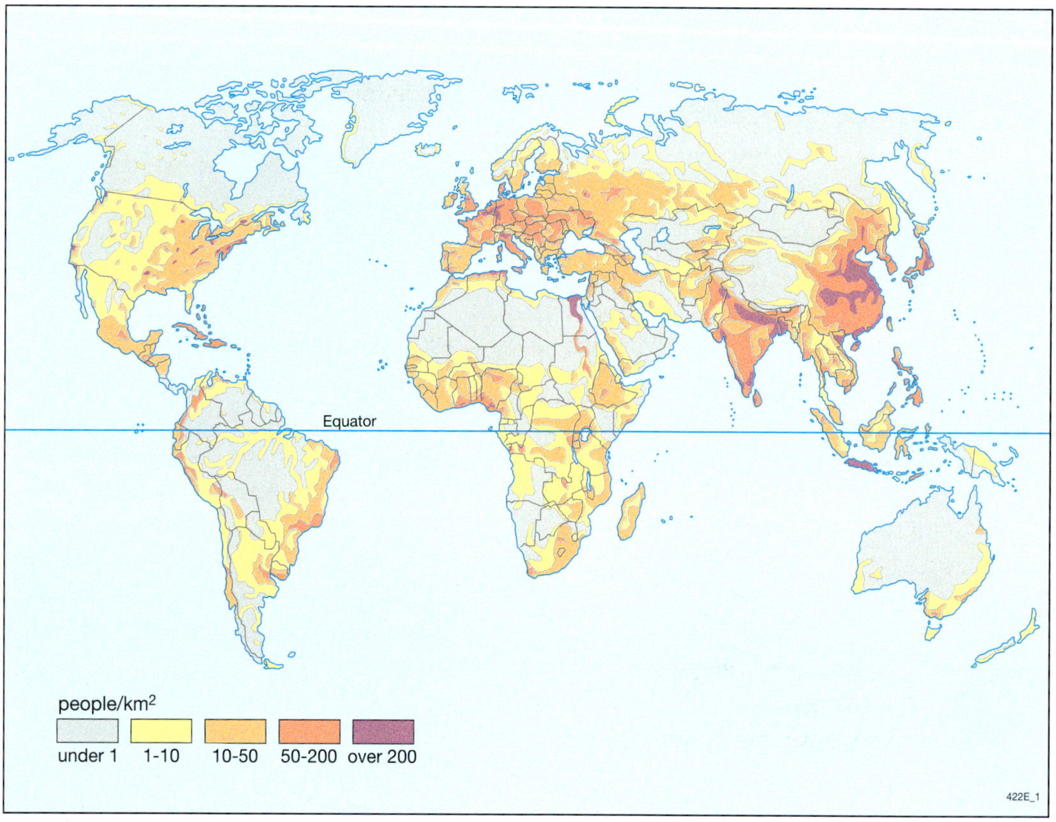


**M2 World population distribution**

## Population density

The distribution of the world population is uneven.
Most people live in Asia, Europe and North America. Here are areas of high population density. Many people live at the same place.

Most Australian people live in the south and southeast. There are no large towns in the yellow area. It has got a low population density.

## Population distribution and density in the U.K.

The population density of the U.K. is 246 persons per square kilometre, but the distribution is not even. You can see that in the thematic map M3.

Towns and areas with industry have got a high population density.

Areas with mountains, forests or farms have got a low population density.

M3 shows with the violet colour that there are six large areas with a population of more than 500 persons per square kilometre in the United Kingdom.

The Scottish Highlands, however, have got a population density of less than one person per square kilometre.

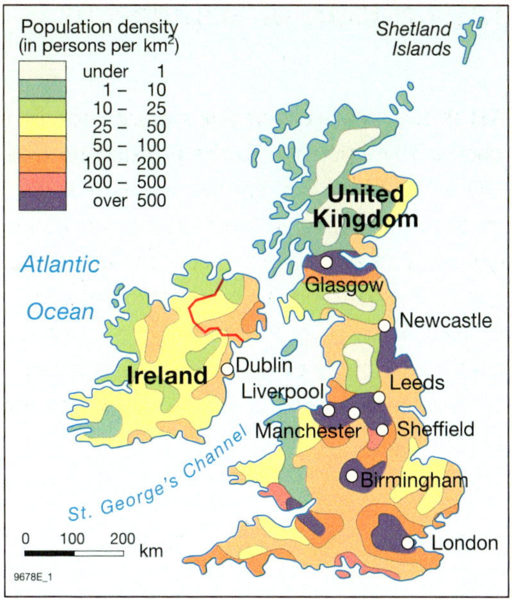


**M3 Population density of the U.K.**

**Europe and the world**

In Europe Monaco has got the highest density with 16,000 persons per square kilometre. Then comes the Netherlands with 482 (Germany 231).

Almost seven million people live in Hong Kong on 165 km$^2$. That is a population density of more than 41,000 inhabitants per square kilometre.

What is the world population at the moment?
www.census.gov/ipc/www/world.html

### Tasks

1. Describe the population distribution in South America.
2. Which areas of Africa are sparsely populated?
3. The population density of Europe is 69 persons per square kilometre. Why is this number not always helpful?
4. List up the names of the areas in the U.K. with a population density of more than 500 persons per square kilometre (Hint: Use the nearest town name).
5. What is the population density in your area?

**Summary**

Six billion people live on Earth.
Every second inhabitant lives in Asia. This continent has also got the highest population density, the two countries with the most people and the most speakers of one language.
Living conditions are not equal for all people.

**Key terms:**
- distribution of population
- population density
- thematic map

## Germany and Europe



**M1 Europe**

# The Federal Republic of Germany

**M1 Germany (Key: page 19)**

## Tasks

**1** Search the map, the text and the fact file for all names that differ from the German names and copy them into your exercise book.
Make a table: cities, states, countries, mountain ranges, rivers, lakes, seas, islands.

Germany has got 16 states. Each state has got its own capital and its own parliament. As you can see on the map M1 some states are huge and others are very small.

1

2

The smallest state is Bremen (1). The largest state, which has also got the highest mountain range, the Bavarian Alps, is Bavaria (2). North Rhine-Westphalia (3) has got the highest number of big cities and inhabitants. In Baden-Wurttemberg (4) you can find Lake Constance, the Black Forest and the source of the River Danube, which are very popular with tourists.

3

4

Rhineland-Palatinate (5) is famous for its many wine regions. The Saarland (6) got its name from the River Saar. In Schleswig-Holstein (7) you need not drive more than 50 km to have a swim in the North Sea or in the Baltic Sea. Lower-Saxony (8) is a favourite place for people who want to watch animals in the Wadden Sea National Park.

5

6

**Key terms:**
- city-state
- government
- parliament
- state

7

8

13

Go to Frankfurt in Hesse (9) to see Germany's highest houses. Thuringia (10), which lies almost in the centre of Germany, is popular for its hilly landscape in the Thuringian Forest.

9

10

14

You can visit famous castles in Brandenburg (11) and Dresden's historical buildings in Saxony (12). Mecklenburg-Western Pomerania (13) is the land of the thousand lakes and the longest German canal runs through Saxony-Anhalt (14). Hamburg (15) has got the biggest sea port of Germany. The home of our government lies in the city-state of Berlin (16).

15

11

12

16

**Germany**
**Population:** 82,541,000 (2003)
**Capital city:** Berlin (3.49 million)
**Land area:** 357,022 km²
**Highest mountain:** Zugspitze (2,962 m)
**Longest river:** Rhine (1,320 km)
**Big cities:** Berlin, Hamburg, Munich, Cologne

# Discover Europe

## The European Continent

Europe is the second smallest of the six continents. It has got over 40 countries with different languages, currencies, and stamps. Its natural borders are the Arctic Ocean, the Ural Mountains, the Caspian Sea, the Caucasus Mountains, the Black Sea, the Mediterranean Sea and the Atlantic Ocean. Some countries have not got a coastline: they are landlocked. Twenty-five of the European countries belong to the European Union (EU).

**Europe**
**Highest mountain:** Mont Blanc (4,804 m)
**Lowest point:** surface of the Caspian Sea (28 m below sea level)
**Biggest city:** Moscow (13,200,000 inhabitants)
**Largest country:** Russia
**Smallest Country:** Vatican City
**Longest river:** Volga (3,600 km)
**Largest lake:** Lake Ladoga (17,700 km²)
**Biggest island:** Great Britain
**Lowest country:** The Netherlands

**Capital**
Every country has got one city that is the capital city. This is the city where the government meets to make decisions. Most are a capital because of past history.
The capital is usally a centre of business and entertainment. This makes most capital cities the largest and most important city in the country.

## Tasks

**1** Take an atlas and find the natural borders of Europe. Then label an outline map with the names of these borders.

**2** List all landlocked countries.

**3** What are the neighbouring countries of … a) Bulgaria, b) Belarus, c) Sweden d) Switzerland. Give the points of the compass.

**Key terms:**
- border (to share a border with)
- capital
- landlocked country
- neighbouring country

# What do Germany's Landscapes Look Like?

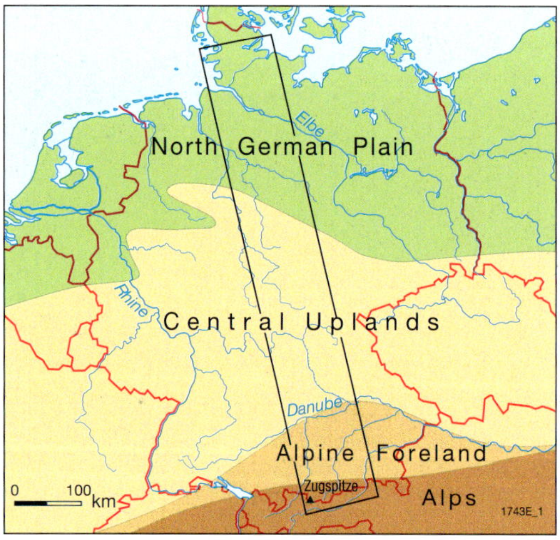


M1 Map of German landscapes

M2 The North German Plain

## Germany's Landscapes

Germany has got three main types of landscapes. In the north there is the North German Plain with a relief that is mostly flat with only some hilly areas.

In the middle part you find the Central Uplands, a hilly and mountainous area with some mountains rising up to about 1,500 m.

The Alpine Foreland and the Alps are in the south. The Alpine Foreland rises slowly from 300 m in the north to 800 m close to the Alps. Finally, the Bavarian Alps have the relief of a steep mountain system with some places higher than 2,000 m.

## Tasks

**1** Describe the photos M2, M4, M5 in your own words.

**2** Where do you want to spend your holidays in Germany? Give reasons.

**Key terms:**
- alpine
- foreland
- hilly area
- plain
- mountainous area
- relief
- upland

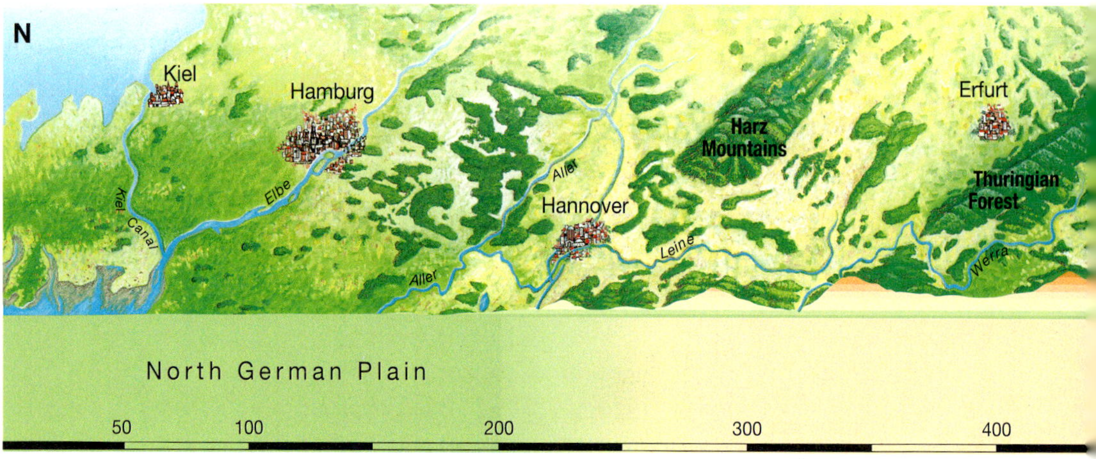


M3 From the coast to the Alps – a cross section from M1

**M4 The Central Uplands**

**M5 The Alpine Foreland and the Alps**

## Summary

Germany has got a lot of different landscapes:

**In the north:**
the North German Plain

**In the middle:**
the Central Uplands

**In the south:**
the Alpine Foreland and the Alps

## European Landscapes and Climate

From Ireland in the west to Russia in the east, from Scandinavia in the north to the Mediterranean Sea in the south, European landscapes are totally different. One reason for this is the climate, which changes throughout the whole continent.
Germany is located in the middle of Europe and has got a transitional type of climate.

**M1 Landscape in Ireland**

Ireland, the 'Green Island' in the west of Europe, has got mild winters and cool summers. So the temperature range is very small. There is a lot of precipitation throughout the whole year. This is typical of a maritime climate.

**M2 A plain in the south of Russia**

Russia is in the east of Europe, far away from the Atlantic Ocean. It has got very warm summers and very cold winters. So the temperature range is very great. There is little precipitation which mostly falls in summer. This is typical of the continental climate of eastern Europe.

**M3 Tundra in the north of Europe**

In the north of Europe there are some areas with long and cold winters and short and cool summers. The summers are relatively wet and the little precipitation in winter falls as snow. This type of climate is called continental climate of northern Europe.

**M4 Spanish coast**

The part of Europe that is close to the Mediterranean Sea has got hot and dry summers and mild and wet winters. It rains a lot in winter because then the winds come from the Atlantic Ocean. This type of climate is called Mediterranean.

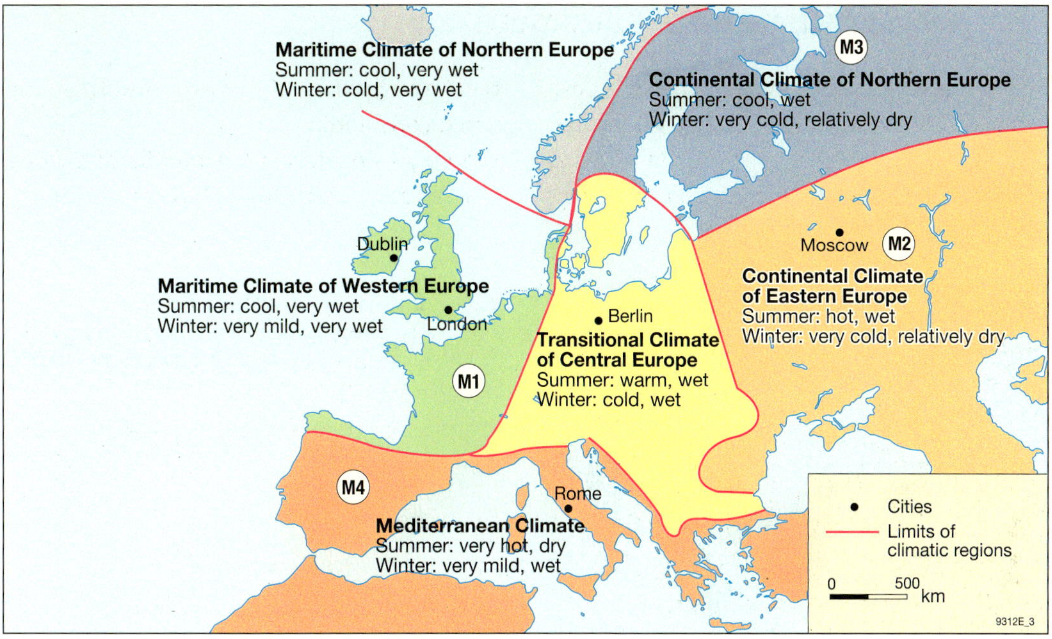


**M5 Climatic Regions of Europe**

Dublin/Republic of Ireland (68 m) 53° 26′ N/6° 15′ W

| | J | F | M | A | M | J | J | A | S | O | N | D | Year |
|---|---|---|---|---|---|---|---|---|---|---|---|---|---|
| °C | 4.5 | 4.8 | 6.5 | 8.4 | 10.5 | 13.5 | 15.0 | 14.8 | 13.1 | 10.5 | 7.2 | 5.8 | 9.6 |
| mm | 71 | 52 | 51 | 43 | 62 | 55 | 66 | 80 | 77 | 68 | 67 | 77 | 769 |

Moscow/Russia (156 m) 55° 45′ N/37° 34′ E

| | J | F | M | A | M | J | J | A | S | O | N | D | Year |
|---|---|---|---|---|---|---|---|---|---|---|---|---|---|
| °C | –10.3 | –9.7 | –5.0 | 3.7 | 11.7 | 15.4 | 17.8 | 15.8 | 10.4 | 4.1 | –2.3 | –8.0 | 3.6 |
| mm | 31 | 28 | 33 | 35 | 52 | 67 | 74 | 74 | 58 | 51 | 36 | 36 | 575 |

**M6 Climate Data**

## Tasks

**1** Compare the climate of Dublin and Moscow as an example. Use the texts and the data of M5 and M6.

**2** How does the climate change from the north of Scandinavia to the south of Italy?

**Key terms:**
- continental climate
- maritime climate
- Mediterranean climate
- temperature range
- transitional climate

**Summary**

Maritime, transitional, continental, and Mediterranean types of climate characterize European landscapes.

# How Can I Draw and Describe a Climate Graph?

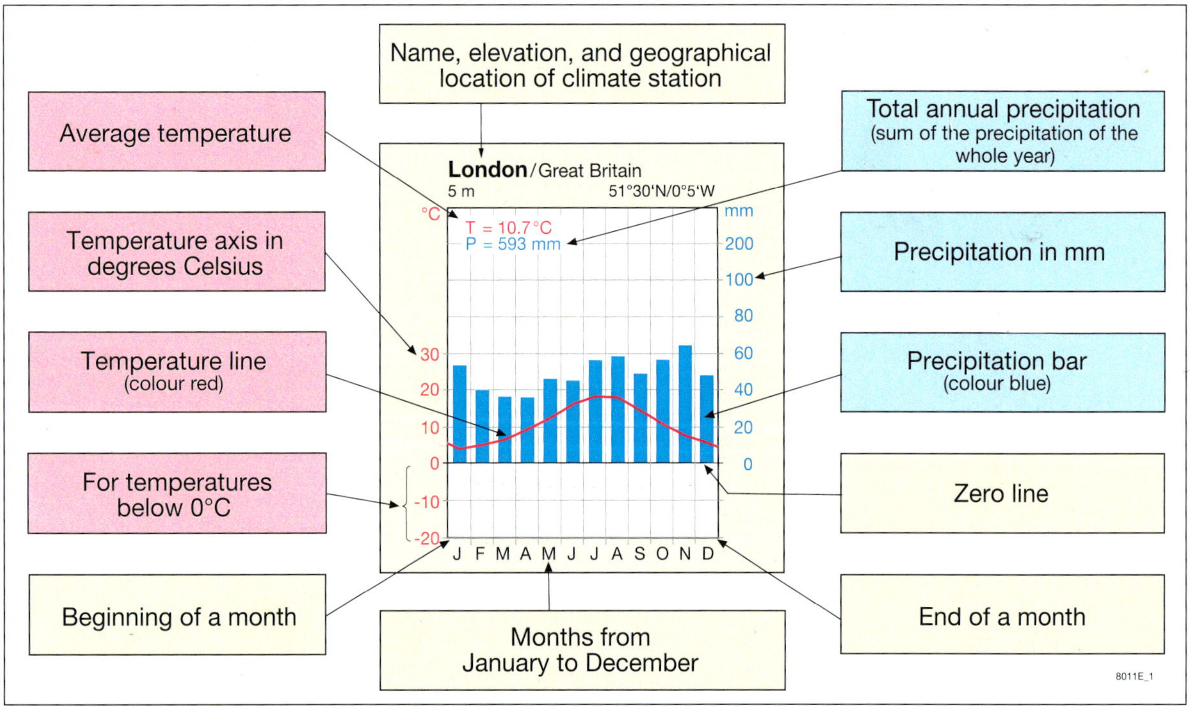


**M1 The parts of a climate graph**

London/United Kingdom (5 m) 51° 30′ N/0° 5′ W

| | J | F | M | A | M | J | J | A | S | O | N | D | Year |
|---|---|---|---|---|---|---|---|---|---|---|---|---|---|
| °C | 4.3 | 5.1 | 6.7 | 9.4 | 12.5 | 16.5 | 17.7 | 17.3 | 14.9 | 11.1 | 7.7 | 5.4 | 10.7 |
| mm | 54 | 40 | 37 | 37 | 46 | 45 | 57 | 59 | 49 | 57 | 64 | 48 | 593 |

**M2 Climate data of London**

**How to draw a climate graph:**

1. Take a sheet of squared paper. Draw the base line and divide it into twelve months (1 month ≙ 1 cm).
2. Draw the left axis for the temperature data (10 °C ≙ 1 cm).
3. Draw the right axis for the precipitation data (20 mm ≙ 1 cm). When there are months with a lot of precipitation choose 20, 40, 60, 80, 100, 200, 300 mm and so on.
4. Now draw the blue bars of precipitation for each month.
5. Then make little dots or crosses in the middle of each month for the values of temperature and connect them with a rounded red line.
6. Write down the name of the climate station, the elevation and the position in the grid system at the top of the climate graph.

**Model text to describe a climate graph:**

1. The climate station is called London. Its location in the grid system is 51° 30′ N and 0° 5′ W. The station is 5 m above sea level and it is located in the south of the United Kingdom.
2. The temperature rises from 4.3 °C in January to 17.7 °C in July. This is a temperature range of 13.4 °C.
3. The driest months are March and April with a precipitation of 37 mm each. The wettest month is November with 64 mm of precipitation. You can say that London has got rainfall throughout the whole year.
4. The average temperature is 10.7 °C. The annual precipitation is 593 mm.

Berlin/Germany (51 m) 52° 28′ N/13° 18′ E

| | J | F | M | A | M | J | J | A | S | O | N | D | Year |
|---|---|---|---|---|---|---|---|---|---|---|---|---|---|
| °C | –0.6 | –0.3 | 3.6 | 8.7 | 13.8 | 17.0 | 18.5 | 17.7 | 13.9 | 8.9 | 4.5 | 1.1 | 8.9 |
| mm | 43 | 40 | 31 | 41 | 46 | 62 | 70 | 68 | 46 | 47 | 46 | 41 | 581 |

Rome/Italy (46 m) 41° 54′ N/12° 29′ E

| | J | F | M | A | M | J | J | A | S | O | N | D | Year |
|---|---|---|---|---|---|---|---|---|---|---|---|---|---|
| °C | 6.9 | 7.7 | 10.8 | 13.9 | 18.1 | 22.1 | 24.7 | 24.5 | 21.1 | 16.4 | 11.7 | 8.5 | 15.5 |
| mm | 76 | 88 | 77 | 72 | 63 | 48 | 14 | 22 | 70 | 128 | 116 | 106 | 880 |

**M3 Climate Data**

## Tasks

**1** Draw a climate graph of Rome (M3) and write a description of the climate graph.

**2** Compare the climate of Rome with that of Berlin with the help of the climate data of M3.

**Key terms:**
- annual precipitation
- average temperature
- axis (pl. axes)
- climate graph
- elevation
- precipitation bar

**Summary**

With the help of three axes, a temperature line and twelve precipitation bars a climate graph shows the climatic situation of a place.

Climate graphs make it easy to compare the climatic situation of different places.

## Oil and Gas in the North Sea – Important Natural Resources

M1 Offshore oil rig

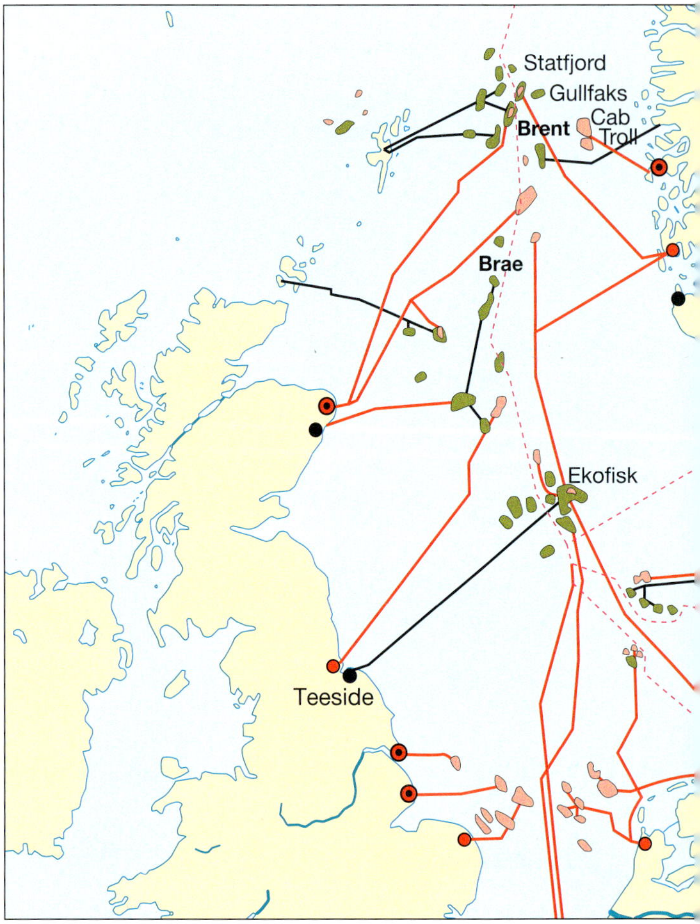


M2 The exploitation of oil and gas in the North Sea

M3 Ship laying a pipeline

M4 Loading a supertanker

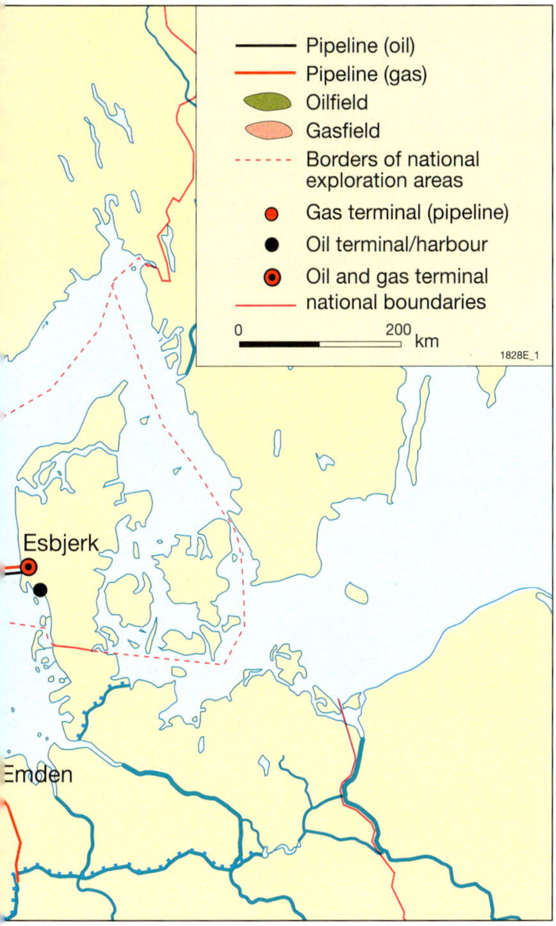


## Oil and gas are natural resources

Oil and gas are natural resources which are limited – or as experts say – non-renewable.

There are oilfields and gasfields deep under the North Sea. You need big oil rigs to drill for oil or gas and to bring it up to the surface.

The companies use pipelines to transport the raw materials to a harbour with an oil or gas terminal. Huge tankers transport the oil or gas to these harbours, too. From there it goes to the nearest oil refinery.

### Tasks

**1** Use the photos (M1, M3, M4, M5) and describe the way of the oil from an oilfield to an oil refinery at the coast.

**2** With the help of M2 and your atlas find out which way ...

a. oil takes from the field Ekofisk to the refinery in Teeside, England,
b. oil takes from the field Brent to the refinery in Esbjerg, Denmark,
c. gas takes from the field Brent to Emden, Germany.

**M5 Unloading tankers at a coastal refinery**

**Key terms:**

| | |
|---|---|
| – to drill for oil (gas) | – oil (gas) field |
| – exploitation of oil (gas) | – oil rig |
| – natural resources | – oil (gas) terminal |
| – non-renewable resources | – pipeline |
| | – refinery |

**Summary**

Oil and gas are very important non-renewable resources.
Companies drill for oil and gas even under the sea and transport it with pipelines and tankers to refineries.

## The Transportation of Oil Can Be Dangerous

M1 Super tanker (up to 400 metres long)

M2 Burning tanker after a collision

M4 Burning oil floating on the sea

M3 Oil rig after an explosion

**If ...**

... an oil tanker has an accident,
... an oil tanker breaks and loses oil,
... the oil starts burning,
... an oil rig is damaged in a storm,
... an oil slick pollutes the water,
... an oil slick reaches the coast,
... a pipeline breaks,
... there is an explosion in an oil refinery,
**... what happens then?**

M5 Removing an oil slick

M6 Oil-covered sea bird

## The effects of an oil spill

After an oil spill a layer of oil covers the surface of the sea. The waves break up the oil slick which forms lumps of oil.

Wind and waves may drive these lumps away from the coast. After some time they sink and affect life in the sea. Fish and plants may get diseases or even die.

When the wind blows the oil to the coast, oil slicks cover the sandy beaches and cliffs. This affects life along the shore, e.g. sea birds and plants may die.

All this, of course, is negative for tourism, too.

1 litre of oil causes the pollution of 1,000,000 litres of water.

## What can we do?

- build bigger tankers, so you need fewer ships
- build oil tankers with double hulls
- let the ships sail on safer routes, e.g. further away from the coast
- better education of captains and their crew
- use more pipelines instead of tankers
- reduce the amount of oil we use
- change from oil to other resources which are less dangerous

### Tasks

**1** Describe how an oil spill affects the life of people, animals and plants.

**2** Explain the solutions in the list above.

**Key terms:**
- disease
- layer of oil
- lumps of oil
- oil slick
- oil spill
- pollution
- (super) tanker

**Summary**

There are many dangers when you transport oil. Accidents affect the life of people, animals and plants. There is no easy solution.

# Agriculture in Great Britain

### M1 Hill farming

**Climate:** Cool summers follow cold winters. Plenty of rainfall is good for the growth of grass.
**Soils:** The thin, poor soils are good enough for sheep grazing but not for other forms of farming.
**Relief:** The hilly, steep land is not suitable for machines or cattle.
**Products:** The sheep give wool and meat.

### M2 Dairy farming

**Climate:** Warm summers and mild winters mean that cattle can be outside all year long. The high rainfall is good for the fast growth of grass.
**Soils:** They are fertile enough for grass growth, but not for arable farming.
**Relief:** We find only gentle slopes. Cattle cannot graze on land that is too steep.
**Products:** The cows produce milk, and, at the end of their lives, meat.

**M3 Main farming types in Great Britai**

**Soil:** Most plants need soil to grow. They get their water, minerals and even the air to breathe from the soil. The plants hold on to the soil with their roots.

**Key terms:**
- arable farming
- dairy farming
- farming type
- hill farming
- market gardening
- soil

nd Ireland

## Tasks

**1** Which farming type depends mainly on a. climate, b. soils, c. relief?

**2** In which farming type do you need: a. most workers? b. most machines?

**Summary**

In Great Britain there are different types of farming. They depend mainly on the climate, the soil and the relief.

**M4 Arable farming**

**Climate:** Warm summers follow frosty winters. The yearly average temperature is 8 °C or more. There is little rainfall during the growing season.
**Soils:** The fertile soils are easy to plough.
**Relief:** It is mostly flat, so it is easy to use machines.
**Products:** The main crops are cereals, like wheat, and root crops, like potatoes.

**M5 Market gardening**

**Climate:** In huge greenhouses, farmers can control water and temperature. The climate outside is not important.
**Soils:** Farmers improve the soil quality with fertilizers.
**Relief:** Greenhouses need flat land.
**Products:** They grow vegetables and flowers.

# Farming for Oilseed Rape, a Renewable Resource

**M1 Farmer Fairs**

This is Mr. Fairs, a farmer in East Anglia. His farm is an arable farm. When Lisa was on holiday in England, she interviewed him.

**Lisa:** *Mr. Fairs, how big is your farm?*
**Mr. Fairs:** That is difficult to say, because we farm not only our farm with 750 acres. We work for ten more farmers and do all the work on the fields for them: we plough, we sow, we harvest. This is "contract farming". Altogether we farm 4400 acres.
**Lisa:** *That's a lot. In Germany only one in fifteen farms is bigger than 250 acres! So you need many workers?*
**Mr. Fairs:** No, we don't. We use machines to do the work for us. So my son and I have got only 3 permanent staff and 3 more staff at harvest time.
**Lisa:** *What do you grow on your farm?*
**Mr. Fairs:** We grow mostly wheat, but also peas and beans. All these crops will become animal feed. And we grow oilseed rape.
**Lisa:** *Are you a farmer or a businessman?*
**Mr. Fairs:** Well, I'm both. This type of farming is agribusiness which means a business in agriculture. I spend most of the time in the office, my son looks after the machines and is out in the fields.
**Lisa:** Thank you, Mr. Fairs!

**M2 These are the crops, farmer Fairs grows in contract farming:**

**wheat**

**peas**

**beans**

**M3 Oilseed rape field in May**

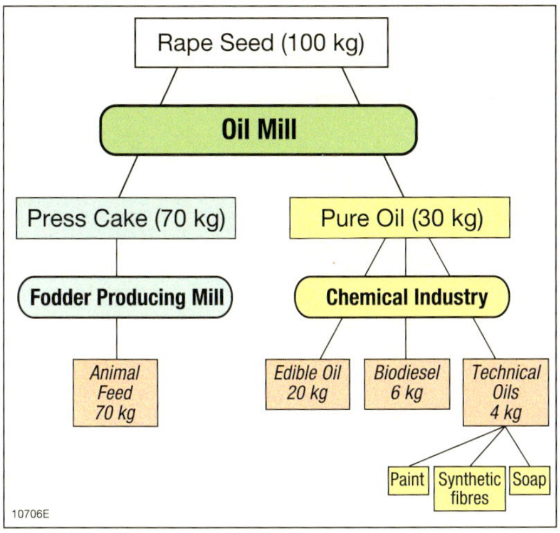


**M4 Products from oilseed rape**

Over the last few years, the area where farmer Fairs has grown oilseed rape has increased to more than one quarter of all the fields he farms today.

"*You get a good price for rape nowadays. There are so many things you can make from the seeds! And they say that from seven square metres of your field you can harvest so many seeds that you can produce 1 litre of oil. The government tells us to grow more of it because it is a renewable resource. This means that you can grow new oilseed rape every year! We farmers can become the oil sheiks of the future*", he smiles.

More and more people now use biodiesel as fuel for their cars because they think it is environmentally friendly and saves the natural resources of oil.

## Tasks

**1** Write a short summary of the interview and the text on page 57.

**2** What are the advantages of oilseed rape farming?

**3** Describe the steps of biodiesel production from the field to the petrol station.

**M5 From field to car**

**Farmer Fairs harvests oilseed rape**

**Oil mill**

**In a chemical plant**

**Biodiesel fuel at the petrol station**

**Key terms:**

- agribusiness
- animal feed
- biodiesel
- contract farming
- acres (0.4 hectare)
- oilseed rape
- renewable resource

# Industry in the London Area

**M1** The **textile industry** produces yarn, fabric and clothes.

**M2** The **metal-producing industry** produces the metals for the metal industry.

**Economic Map**

Economic maps show where you can find
- good soils for agriculture,
- raw materials for mining,
- important industries,
- major transport routes for industrial goods,
- big harbours,
- tourist areas,
- service industries.

So economic maps have got a lot of information for the geographer.

**Key terms:**
- car industry
- chemical industry
- economic map
- electronics industry
- food industry
- iron-producing industry
- metal industry
- raw material
- service industries
- textile industry

## Tasks

**1** Use an economic map in your atlas to find out about the industries around Berlin, Warsaw, and Madrid. Present your results in a table.

**M3 Service industries** are offices, shops, banks, hospitals, and all the places where people serve you.

**M4** The **electronics industry** produces high-tech goods like computers, mobile telephones, and TV sets.

**M5** The **car industry** produces cars and lorries or their components.

**M6** The **chemical industry** produces e.g. plastic, paint, fertilizer, medicine.

**M7 Economic map**

**M8** A **refinery** is at the end of pipelines or in harbours. It uses crude oil to produce petrol, heating oil and raw materials for the chemical industry.

**M9** The **food industry** produces nearly everything we eat and drink.

**M10** The **metal industry** produces everything made of metal, like machines.

# A European Aeroplane for the World

Flying by aeroplane is safe and quick. Nowadays more people use aeroplanes than ever before. They fly on business trips or go on holiday. So the airlines need more aeroplanes.
The world's biggest maker of aeroplanes is a European company, Airbus S.A.S. It has got 16 plants in four different countries in Western Europe (M3). Nearly 50,000 people work for Airbus in Great Britain, France, Spain and Germany (M2) and more than 100,000 in nearly 100 supply companies, most of them in Western Europe and the United States.
Supply companies produce parts for the aeroplanes, for example, Rolls-Royce in England, which makes the engines. The supply companies then send the parts to one of the Airbus plants. Here they use all the components to make parts for the aeroplane, for example the cockpit. All aeroplane parts come together in Toulouse in southern France for the final assembly. From there they fly to Hamburg for furnishing and painting.

http://www.airbus.com/en/

M1 Loading a barge on the Weser River

| Country | Employees |
|---|---|
| F | 19,358 |
| D | 18,423 |
| GB | 8,688 |
| E | 2,726 |
| total | 49,195 |

M2 Employees of Airbus S.A.S. (2003)

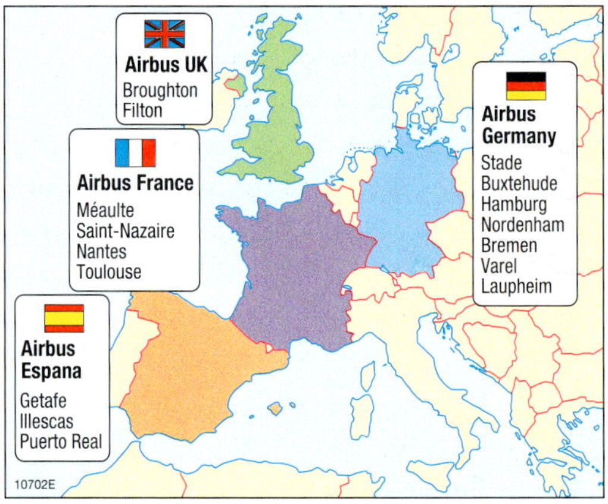


M3 Airbus manufacturing sites in Europe

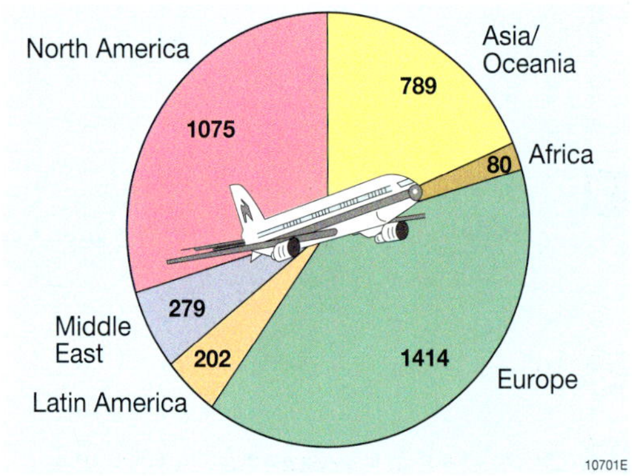


M4 Number of Airbuses all over the world (2005)

**M5 Vessel in the Atlantic Ocean**

**M6 Transport by aeroplane in Hamburg**

## The Airbus A380

The latest child in the Airbus family is the Airbus A380, which is the biggest passenger aeroplane in the world. It can carry up to 800 passengers.
In the photos you can see the different ways to transport aeroplane parts to the final assembly plant in Toulouse. To be able to transport these parts, they had to build new roads, because some of the old ones were not wide enough. They built new ports because the old ports were not deep enough. Then they built new ships, barges to use on rivers and vessels to use at sea. Last but not least they had to build special trucks to transport the big parts on the road.

**Summary**

Many people in Europe work together to build high-tech Airbuses.

**M7 Truck transport in southern France**

### Tasks

**1** Locate the seven Airbus sites in Germany (M3) in your atlas. Show in a table in which states they are.

**2** Describe M3 and M4.

**3** Describe the transport of aeroplane parts on two of the photos.

**Key terms:**
- aeroplane
- barge
- company
- employee
- final assembly
- manufacturing site
- port
- supply company
- truck
- vessel

# Edinburgh, the Capital of Scotland

Edinburgh is the capital of Scotland. It is the seventh largest town in the UK with almost 450,000 inhabitants (M1).

M3 Shopping in Princes Street

## The inner city – Old Town

The town began with Edinburgh Castle in the 11th century (M4). It is on Castle Rock in the southern part of the inner city. From the castle you can see to the Firth of Forth and the Scottish Highlands.
Old Town grew down the hill from the castle to the east and south. The main street, the Royal Mile, is very narrow.

M4 Edinburgh Castle in Old Town

M2 Living in New Town

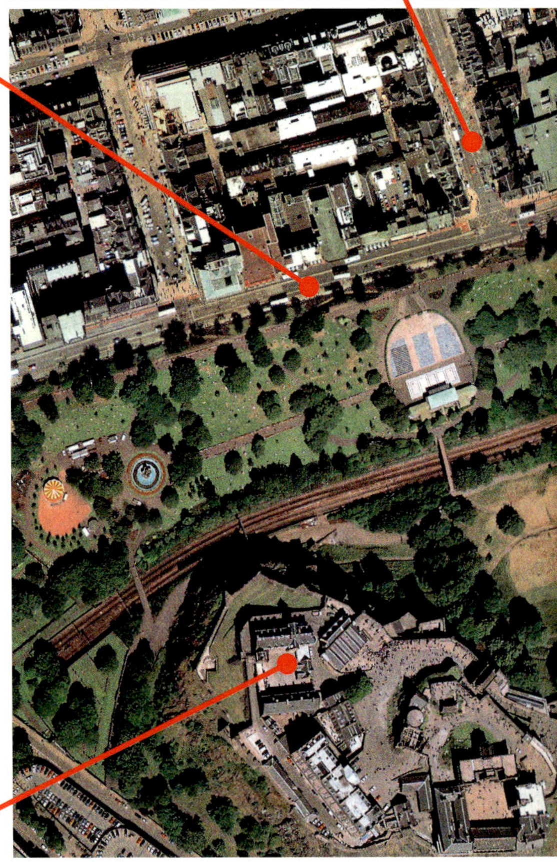

M1 Aerial photograph of Edinburgh's Inner city

## Working and living in Old Town

The Royal Mile is where many Edinburgh people live and work. There are tourist shops, pubs and restaurants (M6, M7). Here you can also find many large government buildings and churches. With all the cars and tourists it is noisy in Old Town for the people who live there.

## The inner city – New Town

Many of the 2,000,000 yearly tourists to Edinburgh arrive at Waverly Station (M5). It is in Princes Garden between Old Town and New Town.

New Town with its straight, planned streets is from the 18th century. Here you can find homes, hotels, offices and shops (M2, M3).

**M5 Waverly Station**

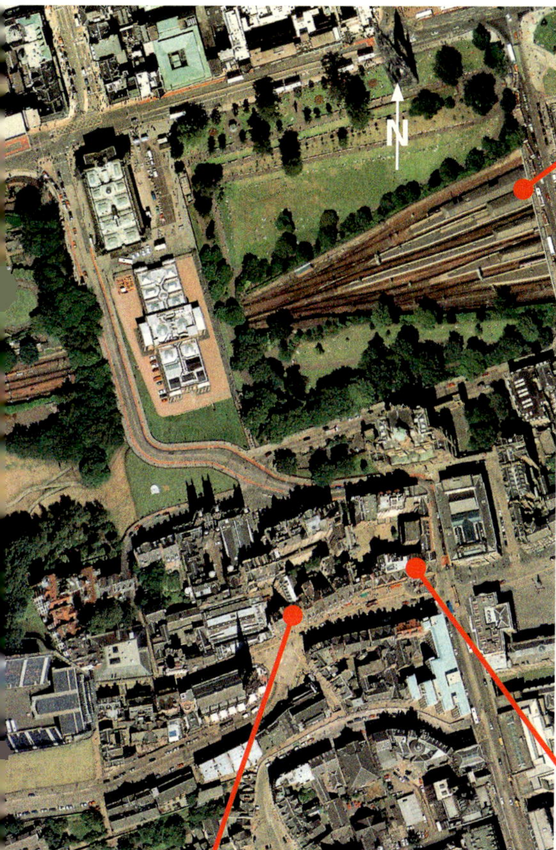


### Tasks

**1** Use the aerial photograph to describe the location of the places (M2-M7).

**2** Compare Old Town and New Town: age, buildings, streets and functions.

**Key terms:**
- aerial photograph
- capital
- government
- the inner city
- tourist

**M7 Deacon Brody's Tavern in the Royal Mile**

**M6 Living in Old Town**

## The outer city

The rest of Edinburgh surrounds Old Town and New Town in the north, south and west (M7).

In the other five parts of town you find residential areas with homes, schools, industry and supermarkets for the larger part of Edinburgh's population.

In the outer city people have more room and newer buildings (M1, M2).

Many jobs are in the north on the Firth of Forth (M8) or at shopping centres (M4).

**M1 Living in the outer city**

**M2 Living in the outer city**

**M3 Driving into the inner city**

**M4 Shopping centre in the outer city**

**M5 View of Arthur's Seat**

M6 Golf – a popular leisure activity

## Leisure in and round Edinburgh

The people of Edinburgh like to relax in the the parks of Old Town and New Town and visit cultural institutions and activities in their leisure time.

They and tourists also enjoy the wonderful view from the top of Arthur's Seat (822 feet high) (M5). When you leave town to the north, there are many possibilities for leisure, especially water activities on the Forth and golf (M6).

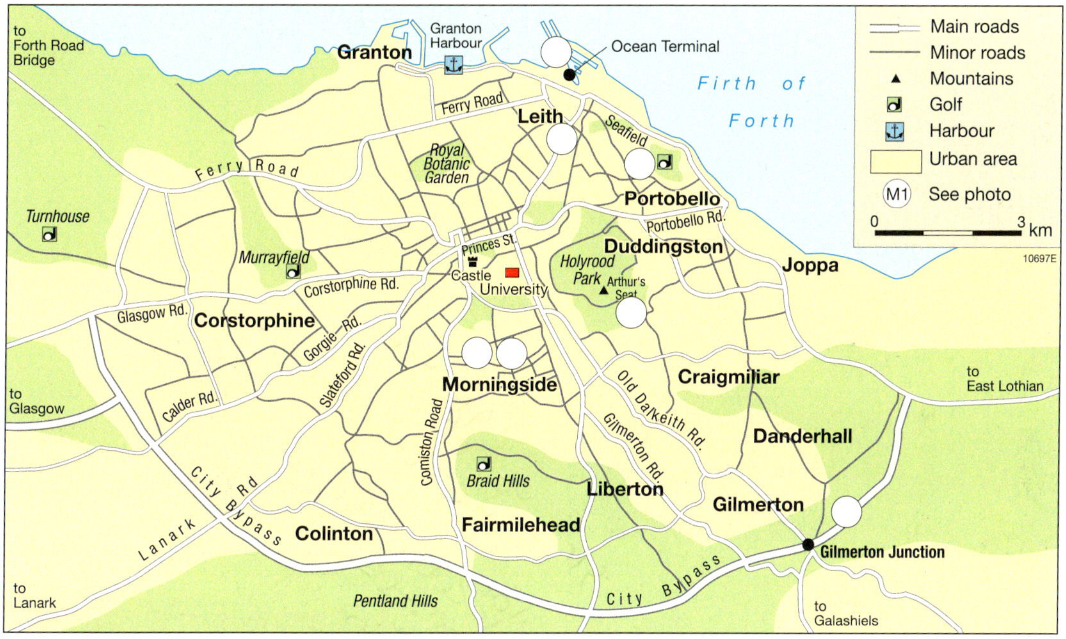


M7 Edinburgh's street network

M8 People work at the Ocean Terminal

**Key terms:**

- industry
- leisure
- the outer city
- residential area

**Tasks:**

**1** Compare the living conditions in the inner city and outer city.

**Summary**

Edinburgh, the capital of Scotland, has got two main parts: the inner city and the outer city. Most people live and work in the outer city.

# The Two Sides of Tourism

### What is Tourism?

When you spend your holidays away from home, you are one of about 680 million tourists all around the world.

Travelling has become easy because cars, trains and planes allow people to reach regions far away.

Travel agencies help tourists to find attractive places all over the world. Package tours include transportation, accomodation and leisure time activities.

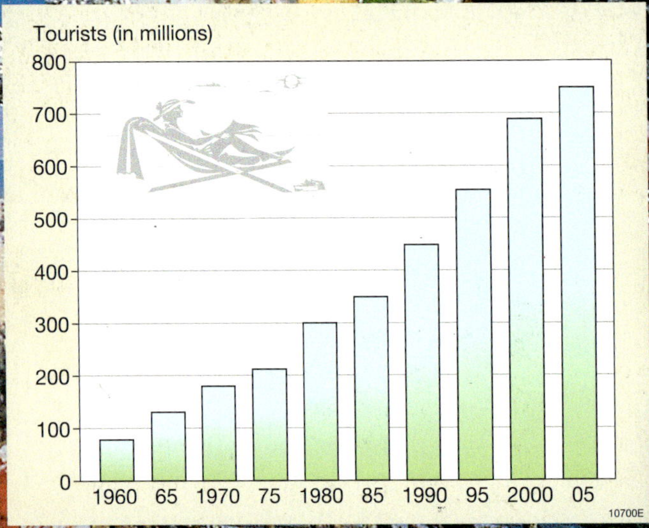


M1 Development of global tourism

## Mass and individual tourism

More and more people use the service of travel agencies to travel to very popular places. This has led to mass tourism which has become an important service industry with many jobs.

Individual tourism, however, means that people plan their holidays on their own to avoid popular places of mass tourism.

These two kinds of tourism affect the environment in different ways:

Mass tourism means that many people want to enjoy their holidays in the same place at the same time. They all need accomodation, restaurants and expect leisure time activities.

Individual tourism does not affect a region that much because there are fewer tourists with fewer demands.

### Tasks

**1** Make a table which shows the two types of tourism and the effects on a tourist region.

**2** Use your atlas and find examples of popular places of mass tourism for summer and winter holidays.

**Summary**

The tourist industry (mass and individual tourism) has become an important economic factor.
Mass tourism has a greater effect on a tourist region than individual tourism.

**Key terms:**
- accomodation
- leisure time activities
- package tour
- mass/individual tourism
- tourism
- travel agency

## Welcome to Malta – a Famous Tourist Place

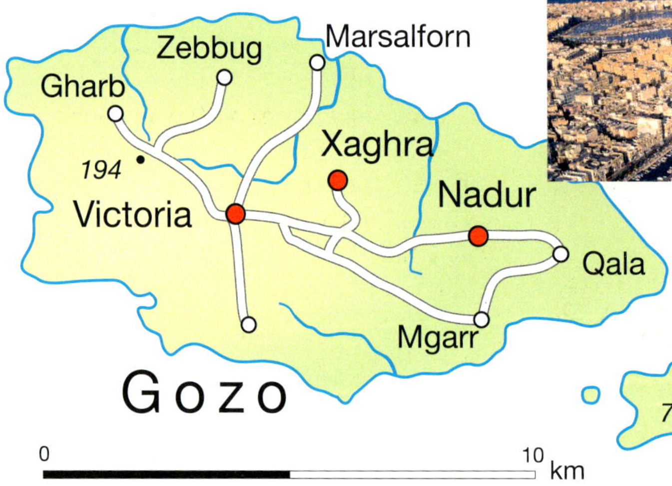

**M1 Harbour with fishing boats and yachts**

**M2 Historical site**

**M3 Typical tourist hotel**

**M4 Climate data Malta**

| | J | F | M | A | M | J | J | A | S | O | N | D |
|---|---|---|---|---|---|---|---|---|---|---|---|---|
| **T h** | 18.1 | 18.0 | 19.5 | 21.8 | 23.1 | 30.4 | 33.2 | 33.6 | 30.8 | 26.8 | 23.0 | 19.8 |
| **T l** | 9.5 | 9.3 | 10.2 | 11.9 | 15.1 | 18.6 | 21.2 | 21.8 | 20.4 | 17.2 | 13.9 | 11.1 |
| **S** | 5.0 | 6.0 | 7.0 | 8.0 | 10.0 | 11.0 | 12.0 | 11.0 | 9.0 | 7.0 | 6.0 | 5.0 |
| **W** | 14.5 | 14.5 | 14.5 | 16.1 | 18.4 | 21.1 | 24.5 | 25.6 | 25.0 | 22.2 | 19.3 | 16.7 |
| **P** | 90.1 | 60.8 | 44.7 | 24.0 | 8.9 | 3.9 | 0.9 | 0.8 | 27.8 | 44.7 | 76.8 | 100.2 |

T h Highest day temperature (°C)
T l Lowest day temperature (°C)
S Hours of sunshine
W Water temperature (°C)
P Precipitation (mm)

M5 Diving off the coast of Malta

M6 Arch formed by the sea

**Malta**

The Republic of Malta consists of several islands. The biggest three are Malta, Gozo and Comino. Spain, Italy and Arab countries played an important role in the early history of Malta. From 1814 until 1964 it was a British colony. Today it is an independent republic within the British Commonwealth.

| | |
|---|---|
| Total area: | 316 km$^2$ |
| Population: | 402,000 (2004) |
| Capital: | Valletta (7,200 inhabitants) |
| Language: | Maltese, English, Italian |
| EU member: | since 2004 |
| Tourism: | 33 % of all jobs are in the tourist industry. |

## Tasks

**1** Describe the location of Malta and find out how you can get there.

**2** Use M1–M6 and explain why so many tourists go to Malta for their holidays.

**3** Look at the reasons you have found in task 2. Use M4 and explain which time of the year is best for tourists.

**Key terms:**
- holiday destination
- historical site
- tourist activities

**Summary**

Malta is an island in the Mediterranean Sea which has become a famous holiday destination. The main reasons are the Mediterranean climate and landscape, as well as historical places and many different kinds of leisure time activities.

## On the Way to a United Europe

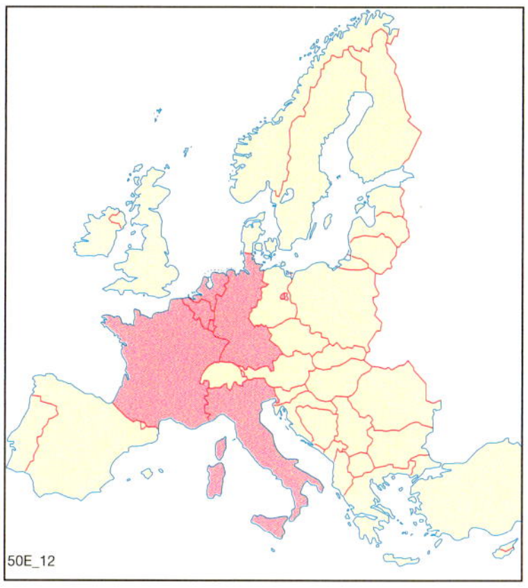

**M1** In 1957 these six countries formed the European Economic Community (EEC), which in 1967 became the European Community (EC).

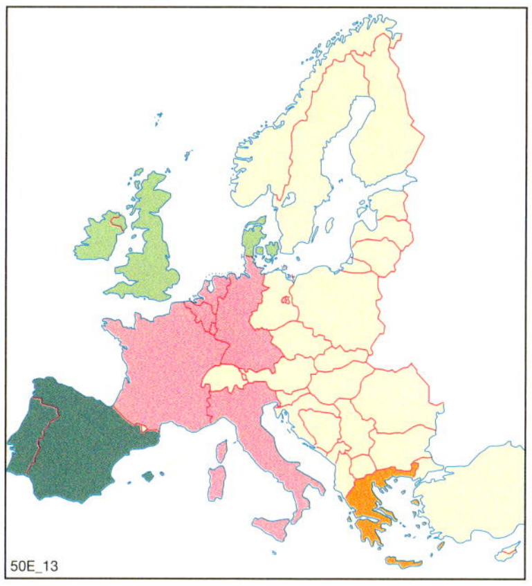

**M2** In 1973 Great Britain, Ireland and Denmark joined the EC. In 1981 Greece and in 1986 Spain and Portugal followed.

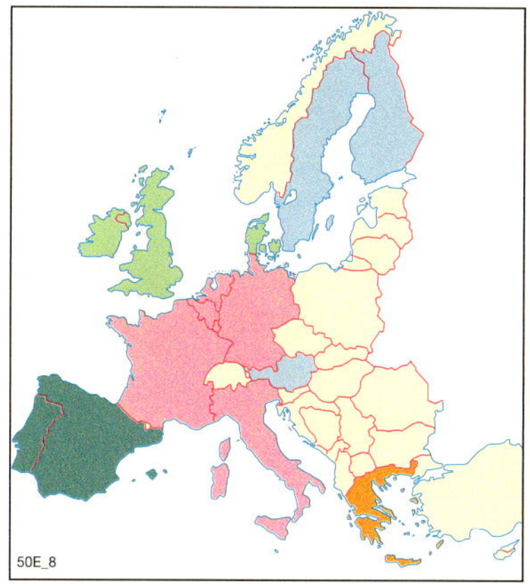

**M3** In 1993 the EC became the European Union (EU). Austria, Finland and Sweden joined the EU in 1995.

Today there are 25 member states in the EU. Each of them has got a government but there is no European Government. There are four main institutions instead:

- **The European Council**
  The leaders of the EU states meet and discuss what is best for the EU.
- **The European Commission**
  Each member state sends politicians who together propose new laws for the EU.
- **The European Parliament**
  The members discuss the proposals of the European Commission and decide on them.
- **The European Court of Justice**
  When there are problems, this court of law tries to solve them.

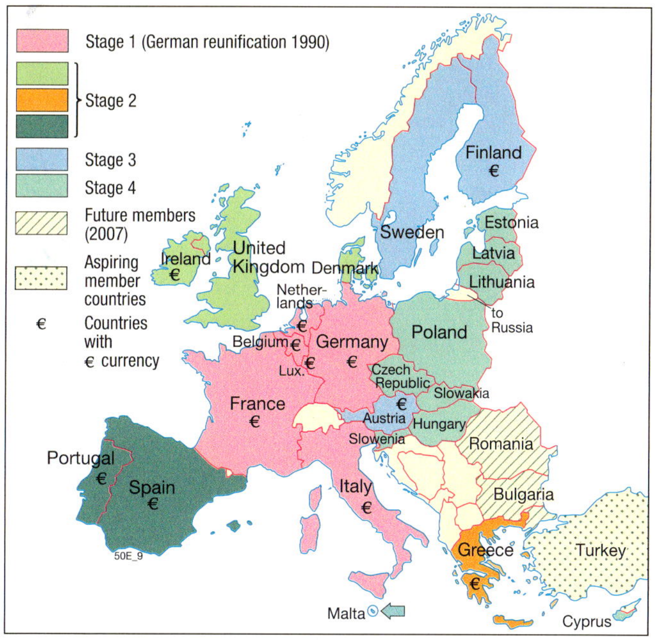


**M4** In 2004 ten more countries joined the EU: Cyprus, the Czech Republic, Estonia, Hungary, Latvia, Lithuania, Malta, Poland, Slovakia and Slovenia.

**Key terms:**
- European Commission
- European Court of Justice
- European Council
- European Parliament
- European Union (EU)

## Tasks

**1** Describe the development of the EU since 1957.

**2** Find out which other European countries want to join the EU in future.

**3** Match the flags and countries in M4 (use the Internet).

# Living Together in Europe

**M1 Open borders**

## Crossing borders has become easy

When people went through passport controls in Europe in former times, they had to show their passports and go through customs. Sometimes this could take quite a long time.

The Netherlands, France, Belgium, Luxembourg and Germany were the first European countries where there were no checks at the border any longer. Today this has become quite normal in many European countries and sometimes people don't even notice that they are passing a border.

**M3 Getting across Europe**

**M2 The Euro replaces national currencies**

## Driving in Europe has become easy

The European Union has introduced common documents, such as ID-cards, passports, driving licenses and number plates. Control is very easy because the police and other authorities in European member states understand the documents at once.

## Shopping in Europe has become easy

People who travel to another country no longer have to exchange money because they can use the European currency, the Euro, in 12 member states (page 71 M4). In addition, it is easier to compare prices.

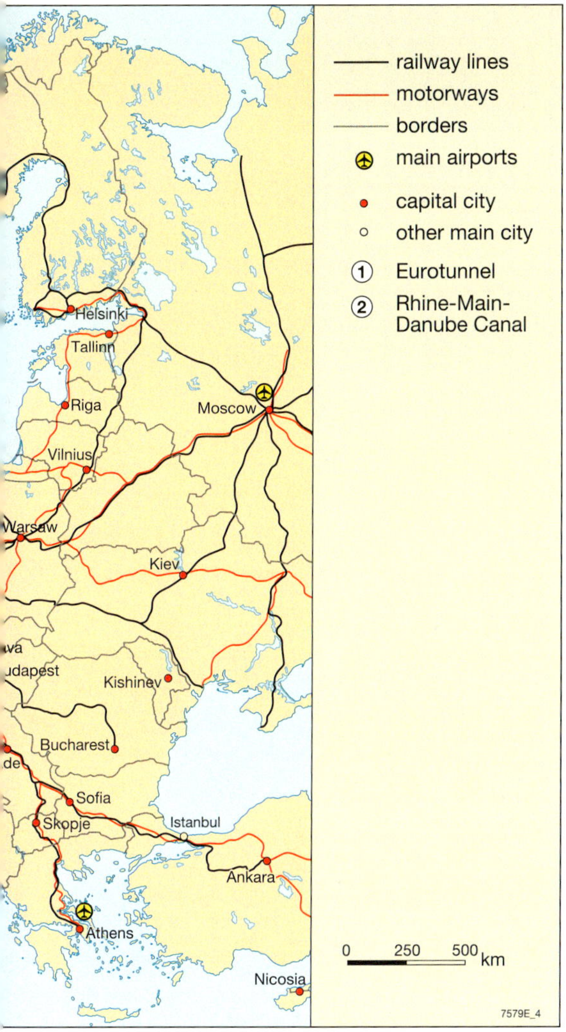

**M4 The Thalys – a high-speed train**

## Travelling in Europe has become easy

High speed trains link many big European cities. The Thalys, for example, runs from Cologne to Paris via Brussels. Here people can change trains and travel to London using the Eurostar.

A motorway network links nearly all European countries, too. The E 45, for example, runs from Arhus (Denmark) to Bologna (Italy) via Munich.

Airlines offer flights from even regional airports like Hahn/Frankfurt to destinations all over Europe, for example, to Manchester or Reus/Barcelona.

Waterways, such as the Rhine-Main-Danube Canal, link the Europort of Rotterdam with central and south-eastern Europe.

**Key terms:**
- common currency
- passport control
- customs
- European transportation network

**M5 The new European number plate**

**Summary**

Living together has become easier in Europe because of modern and fast transportation networks, passing borders without any controls and using the common currency, the Euro.

## Tasks

**1** Choose different routes across Europe and find out whether you have to show your ID-card and/or have to exchange money.

**2** Travelling in Europe has become easy. Explain.

**Aborigines** australische Ureinwohner
**accident** Unglück
**accommodation** Unterkunft
**activity** Aktivität
**aerial photograph** Luftbild
**aeroplane** Flugzeug
**affect (to)** beeinträchtigen, beeinflussen
**agree with (to)** übereinstimmen mit
**agribusiness** Agrobusiness
**agriculture** Landwirtschaft
**air pressure** Luftdruck
**airline** Fluggesellschaft
**allow (to)** ermöglichen, erlauben
**all year long** das ganze Jahr über
**almost** fast
**alpine** alpin, Alpen-
**alternative** Alternative
**although** obwohl
**amount (of)** Menge (von)
**ancient** alt
**anemometer** Windgeschwindigkeitsmesser
**angle** Winkel
**animal feed** Tierfutter
**annual** jährlich
**annual precipitation** Jahresniederschlag
**apart from** außer
**app. (approximately)** ungefähr (ca.)
**arable** Acker-
**arable farming** Ackerbaubetrieb
**arch** Bogen
**Arctic Circle** Nordpolarkreis
**area** Fläche, Gebiet
**arrival** Ankunft
**arrow** Pfeil
**aspire** anstreben, aufstreben
**as the crow flies** Luftlinie
**astronaut** Astronaut
**at the same time** zur selben Zeit
**atmosphere** Atmosphäre
**attract (to)** anziehen
**attractive** attraktiv
**Austrian** Österreicher/in
**authority** Behörde
**average** Durchschnitt, durchschnittlich
**average temperature** Durchschnittstemperatur
**avoid (to)** vermeiden
**axis** Achse

**background** Hintergrund
**balloon** Luftballon
**bar** Säule (in einer Grafik)
**bar graph** Säulen-, Balkendiagramm
**barge** (Fluss)Schiff
**barometer** Luftdruckmesser
**barrel** Fass (Maßeinheit, 159 Liter)
**bay** Bucht
**beach** Strand
**bean** Bohne
**bear** Bär
**Belgian** Belgier/in
**biodiesel** Biodiesel
**birth rate** Geburtsrate
**blank** leer
**blizzard** Eissturm, Kältesturm
**blow (to)** blasen hier: treiben
**boil (to)** kochen
**border** Grenze
**bottom half** untere Hälfte
**brainstorming** Brainstorming (Ideensammlung)
**brass paper fastener** Versandbeutelklammer
**break (to)** zerbrechen
**breathe (to)** atmen
**British** Brite/Britin
**brush** Pinsel
**built in** eingebaut
**business** Wirtschaft, Geschäft
**business trip** Geschäftsreise
**bypass road** Umgehungsstraße

**call (to)** hier: heißen
**capital** Hauptstadt
**cardboard** Karton
**car industry** Automobilindustrie
**carpet** Teppich
**cart** Fuhrwerk, Karren
**case** Fall
**castle** Schloss, Burg
**caterpillar** Raupe, Planierraupe
**cathedral** Kirche (Bischofskirche)
**cattle** Rindvieh
**cause (to)** verursachen
**century** Jahrhundert
**certain** bestimmte
**certain point** bestimmter Punkt
**CET (Central European Time)** MEZ (Mitteleuropäische Zeit)
**characteristic** typisch
**chart** Schaubild, Diagramm
**check** Kontrolle
**check (to)** kontrollieren
**chemical industry** chemische Industrie
**Christianity** Christentum
**church** Kirche
**circle** Kreis
**circle (to)** umkreisen
**city, cities** Stadt, Städte
**city-state** Stadtstaat
**cliff** Steilküste
**climate** Klima
**climate data** Klimadaten
**climate graph** Klimadiagramm
**clothes** Kleidung
**cloud** Wolke
**cloud cover** Wolkenbedeckung
**coal** Kohle
**coast** Küste
**coastline** Küstenlinie
**coat of arms** Wappen
**code** Code (Kennbuchstaben)
**coin** Münze

**cold** kalt
**collect (to)** sammeln
**collision** Kollision, Unfall
**colony** Kolonie
**comet** Komet
**common** gemeinsam
**common currency** Gemeinschaftswährung
**company** Firma, Unternehmen
**compare (to)** vergleichen
**compass** Kompass
**competition** Wettbewerb
**complete** vervollständigen
**component** Zubehörteil
**condition** Bedingung
**conflict** Konflikt
**connect (to)** verbinden
**connected to** verbunden mit
**consist of (to)** bestehen aus
**consumer** Verbraucher
**contain (to)** beinhalten
**continent** Kontinent
**continental climate** Kontinentalklima
**contract farming** Vertragsanbau
**convert (to)** konvertieren, umrechnen
**Council of Ministers** Ministerrat
**country** Land, Staat
**countryside** ländliche Gegend
**Court of Justice** Gerichtshof
**cover (to)** bedecken
**crate** Getränkekiste (Plastik)
**create (to)** (er)schaffen
**crew** Mannschaft
**crop** Feldfrucht
**cross (to)** überqueren
**cross out (to)** durchstreichen
**cross section** Querschnitt, Profil
**crude oil** Erdöl, Rohöl
**cube** Würfel
**cubic** Kubik-
**cultural institution** Kultureinrichtung
**culture** Kultur
**currency** Währung
**current** Strömung
**curved** hier: gewölbt
**customer** Kunde
**customs** Zoll
**Cypriot** Zypriote/Zypriotin
**Czech** Tscheche/Tschechin

**dairy farm** Milchbauernhof
**dairy farming** Milchwirtschaft
**damage** Schaden
**damaged** beschädigt
**Dane** Däne/Dänin
**data** Daten, statistische Werte
**daytime** Tageszeit, Tageslicht
**dateline** Datumsgrenze, Datumswechsel
**decide (to)** entscheiden
**decimal** Dezimalzahl
**deep** tief
**degree** Grad
**demand** Anspruch
**dense** dicht
**depth** Tiefe
**describe (to)** beschreiben
**desert** Wüste
**destination** Ziel(ort)
**diameter** Durchmesser
**die (to)** sterben
**differ from (to)** sich unterscheiden von
**difference** Unterschied
**different** verschieden
**direct (to)** hier: führen
**direction** Richtung, hier: Himmelsrichtung
**discuss (to)** diskutieren, beraten, besprechen
**disease** Krankheit
**disk** Scheibe
**display** hier: Bildschirm
**distance** Entfernung
**distribution** Verteilung
**distribution of population** Bevölkerungsverteilung
**dock** Hafenanlage
**document** Dokument, Ausweis
**dot** Punkt
**double** doppelt
**draw (to)** zeichnen, malen
**drill for oil (gas) (to)** nach Öl (Gas) bohren
**driving license** Führerschein
**drizzle** Nieselregen
**droplet** Tröpfchen
**dry (to)** trocknen
**Dutchman/Dutchwoman** Niederländer/in

**each** jede, jeder, jedes
**Earth** Erde
**Earth's axis** Erdachse
**economic map** Wirtschaftskarte
**edible oil** Nahrungsmittel-Öl
**education** Ausbildung, Erziehung
**effect** Folge, Auswirkung
**either ... or** entweder...oder
**electronics industry** Elektroindustrie
**elevation** Höhenlage
**employee** Angestellter
**encyclopaedia** Lexikon
**engine** Triebwerk, Motor
**enough** genug
**entertainment** Unterhaltung, Verknügen
**environment** Umwelt
**environmental** umwelt-
**environmentally friendly** umweltfreundlich
**equal** gleich
**Equator** Äquator
**eroded** erodiert, ausgewaschen
**escape (to)** entkommen
**Estonian** Este/Estin
**EU (European Union)** Europäische Union (Abk.)
**European** europäisch
**European Commission** Europäische Kommission

**European Council** Europäischer Rat
**European Court of Justice** Europäischer Gerichtshof
**European Economic Community** Europäische Wirtschaftsgemeinschaft
**European Parliament** Europäisches Parlament
**even** gleichmäßig
**exchange (to)** wechseln
**exploit (to)** ausbeuten, fördern
**exploitation of** Förderung
**explore (to)** erkunden
**explosion** Explosion

**fabric** Stoff
**face down** mit der Vorderseite nach unten
**facts** Fakten, Tatsachen
**fairy tale** Märchen
**far** fern
**farm (to)** Landwirtschaft betreiben
**farmer** Landwirt, Bauer
**farming** Landwirtschaft
**farming type** landwirtschaftliches Betriebssystem
**fashion** Mode
**feature** Merkmal
**Federal Republic of Germany** Bundesrepublik Deutschland
**female** weiblich
**fertile** fruchtbar
**fertilizer** Dünger
**fewer** weniger
**fibre** Faser
**fill (to)** füllen
**final assembly** Endmontage
**finally** schließlich
**findings** Ergebnisse
**Finn** Finne/Finnin
**fisherman** Fischer
**flat** flach
**flatten (to)** flach machen
**float (to)** treiben
**fodder producing mill** Futtermittelmühle
**fog** Nebel
**fold (to)** falten
**follow (to)** folgen
**food** Nahrungsmittel
**food industry** Nahrungsmittelindustrie
**forecast** Vorhersage
**foreground** Vordergrund
**foreign currency** Devise, ausländische Währung
**foreland** Vorland
**forest** Wald
**form (to)** bilden
**former times** früher
**fraction** Bruchzahl
**freeze (to)** frieren, gefrieren
**Frenchman/Frenchwoman** Franzose/Französin
**frosty** frostig
**frozen** gefroren
**fuel** Treibstoff
**function** Funktion, Zweck
**funnel** Trichter
**furnishings** Einrichtungen, Ausstattung
**further away from** weiter weg von

**galaxy** Sternhaufen
**gap** Lücke
**gas** Erdgas
**gas field** Gasvorkommen, Gaslagerstätte
**gas terminal** Gasanlegestation
**gentle** sanft
**German** Deutscher/Deutsche
**glacier** Gletscher
**global** global, weltweit
**global grid** Gitternetz
**globe** Globus
**goods** Waren, Güter
**government** Regierung
**graph** Diagramm
**grassland** Wiese, Weideland, Grasfläche
**graze (to)** weiden, grasen
**grazing** Weideland
**Greek** griechisch, Grieche/Griechin
**greenhouse** Gewächshaus
**grid system** Gitternetz (von Längen- und Breitengraden)
**ground** Boden
**grow (grew, grown) (to)** wachsen, anbauen
**grow crops (to)** Ackerbau betreiben
**growth** Wachstum
**guess (to)** raten
**guide** Führer

**hail** Hagel
**hailstones** Hagelkörner
**handle (to)** handhaben
**happen (to)** geschehen
**harbour** Hafen
**harvest** Ernte
**harvest (to)** ernten
**heating** Heizung
**heating oil** Heizöl
**heavy rain** starker Regen
**heavy snowfall** starker Schneefall
**hectare** Hektar (100 m x 100 m)
**height** Höhe
**hemisphere** Halbkugel
**herbs** Kräuter
**high-speed train** Hochgeschwindigkeitszug
**high-tech** Hochtechnologie
**hill** Hügel
**hill farm** Bergbauernhof
**hill farming** Bergbauernwirtschaft
**hilly** hügelig
**hilly area** hügeliges Gebiet
**hinterland** Hinterland
**historic** geschichtlich
**historical site** historische Stätte
**hold on (to)** festhalten

**holiday destination** Urlaubsziel
**holiday resort** Ferienanlage
**holy** heilig
**horizontal** waagerecht
**horse** Pferd
**hot** heiß
**huge** groß, riesig
**Hungarian** Ungar/in
**hurricane** Hurrikan (Wirbelsturm)

**ice cap** Eiskappe
**iceberg** Eisberg
**ID-card** Personalausweis
**imaginary** vorgestellt
**important** wichtig
**in addition** zusätzlich, außerdem
**in fact** tatsächlich
**incorrect** falsch
**increase (to)** anwachsen
**independent** unabhängig
**indicate (to)** anzeigen
**individual tourism** Individualtourismus
**industry** Industrie
**inform (to)** (sich) informieren
**inhabitant** Einwohner
**inner city** Innenstadt
**instead (of)** stattdessen, anstatt
**institution** Institut, Einrichtung
**International Date Line** Internationale Datumsgrenze
**introduce (to)** einführen
**invisible** unsichtbar
**Irishman/Irishwoman** Ire/Irin
**iron-producing industry** Eisen verarbeitende Industrie
**island** Insel
**Italian** Italiener/in

**jar** Einmachglas, Konfitüreglas
**job** Arbeitsplatz
**jumbled** durcheinander geschüttelt

**key** Schlüssel, Erklärung (hier: Legende)
**kinds, sorts (of)** Arten (von)
**knitting-needle** Stricknadel

**label (to)** beschriften
**lake** See
**land area** Fläche
**landforms** Landschaftsformen
**landlocked country** Binnenland
**landmass** Landmasse
**landscape** Landschaft
**language** Sprache
**large** groß
**last but not least** nicht zu vergessen
**later on** später
**Latvian** Lette/Lettin
**law** Gesetz
**lay (to)** verlegen
**layer of oil** Ölschicht
**leader** Führer
**left half** linke Hälfte
**leisure (time)** Freizeit
**leisure (time) activities** Freizeitaktivitäten
**length** Länge
**less** weniger
**lie (to)** liegen
**life-boat** Rettungsboot
**like** wie
**limited** begrenzt
**line** Linie
**line graph** Kurvendiagramm
**line of latitude** Breitengrad
**line of longitude** Längengrad
**line up with (to)** hier: sich decken mit
**link (to)** verbinden
**list (to)** nennen, aufzählen
**listed** aufgelistet
**Lithuanian** Litauer/in
**litter (to)** verschmutzen (die Umwelt)
**live (to)** leben
**living conditions** Lebensbedingungen
**load (to)** laden
**local** örtlich, heimisch
**locate (to)** einen Ort bestimmen, ausfindig machen
**location** Lage
**lorry, (pl) lorries** Lastwagen
**lose (to)** verlieren
**lot, a** viel
**lot of, a** viele
**low** niedrig, tief
**lumps (of oil)** (Öl-) Klumpen
**Luxembourger** Luxemburger/in

**magnetic** magnetisch
**mainly** hauptsächlich
**make decisions** Entscheidungen treffen
**male** männlich
**Maltese** Malteser/in
**manufacturing site** Produktionsstätte
**map** Landkarte
**map-makers** Kartenhersteller
**maritime climate** maritimes Klima, Seeklima
**mark (to)** eintragen, markieren, kennzeichnen
**market garden** Gartenbaubetrieb
**market gardening** Gemüse- und Blumenanbau
**mass tourism** Massentourismus
**maximum and minimum thermometer** Maximum-Minimum-Thermometer
**mean (to)** bedeuten
**measure** Maß
**measure (to)** messen
**meat** Fleisch
**medicine** Medizin
**Mediterranean** Mittelmeer
**Mediterranean climate** Mittelmeerklima

**medium** mittelmäßig
**meet (to)** zusammen kommen, sich treffen
**member** Mitglied
**memory** Gedächtnis
**memory stick** Speicherstift, Memory stick
**mentioned** erwähnt
**mercury** Quecksilber
**metal industry** Metallindustrie
**Milky Way** Milchstraße
**mill** Mühle
**mineral** hier: Nährsalz
**mining** Bergbau
**miss (to)** vermissen
**mixed farm** landwirtschaftlicher Gemischtbetrieb
**mobile telephone** Handy
**model** Modell
**moon** Mond
**most of which** von denen die meisten
**mostly** größtenteils
**mountain** Berg, Gebirge
**mountain range** Gebirge, Gebirgskette
**mountainous** gebirgig
**mudflat** Watt, Wattenmeer

**narrow** eng, schmal
**national exploration area** nationales Hoheitsgebiet
**natural resource** Rohstoff
**nearest** nächstgelegen
**nearly** fast
**needed** benötigt werden
**needle** Nadel
**neighbouring** benachbart
**neighbouring country** Nachbarstaat
**network** Netzwerk
**nighttime** Nachtzeit, Dunkelheit
**non-renewable resource** nicht erneuerbarer Rohstoff
**North Pole** Nordpol
**Northern Hemisphere** nördliche Hemisphäre
**north, northern** Norden, nördlich
**notice (to)** bemerken
**nowadays** heutzutage
**number** Anzahl, Zahl
**number (to)** nummerieren

**oblique** schräg
**observe (to)** beobachten
**ocean** Ozean, Meer
**offer (to)** bieten
**office** Büro
**official** offiziell
**offshore oilrig** Ölplattform
**oil field** Ölvorkommen, Öllagerstätte
**oil film** Ölfilm
**oil rig** Bohrinsel
**oil slick** Ölschlamm
**oil spill** Ölunfall, Leck
**oilseed rape** Raps
**oil terminal** Ölanlegestation
**on one's own** selbst
**order** Reihenfolge
**organize (to)** organisieren
**orientated** hier: ausgerichtet nach
**outer city** Außenbezirke einer Stadt
**outline map** Umrisskarte
**overcast** bedeckt
**overcrowded** übervölkert
**overhead** von oben, senkrecht
**own** eigene

**package tour** Pauschalreise
**paint** Farbe
**painter** Maler
**painting** Anstrich
**palace** Palast
**parliament** Parlament, Regierung
**part** Teil
**part of (be)** hier: gehören zu
**passenger** Passagier
**passport** Pass
**passport control** Passkontrolle
**past history** historische Vergangenheit
**pea** Erbse
**peak** Spitze
**peel (to)** schälen
**people** Menschen
**peoples** Völker
**per** pro
**per cent** Prozent
**permanent** ständig
**perish (to)** verderben
**petrol** Benzin
**petrol station** Tankstelle
**physical map** physische Karte
**pie chart** Tortendiagramm, Kreisdiagramm
**pilgrim** Pilger
**pin** Stecknadel
**pipeline** Rohrleitung, Pipeline
**place** Ort
**plain** Ebene
**planet** Planet
**planned** geplant
**plant** Pflanze; Betrieb, Fabrik
**plenty** reichlich
**plough (to)** pflügen
**plywood** Sperrholz
**point** Stelle, Punkt
**points of a compass** Himmelsrichtungen
**point to (to)** zeigen auf
**Pole** Pole/Polin
**politician** Politiker
**pollute (to)** verschmutzen (die Umwelt)
**pollution** Verschmutzung (der Umwelt)
**poor** armselig, wenig ertragreich, kärglich
**popular** beliebt
**population** Bevölkerung
**population density** Bevölkerungsdichte
**port** Hafen

**Portuguese** Portugiese/Portugiesin
**position** Position, Ort
**potato** Kartoffel
**precipitation** Niederschlag
**present (to)** vorstellen, präsentieren
**press cake** Presskuchen
**pressure** Druck
**Prime Meridian** Meridian, Längengrad 0
**printed** gedruckt
**produce (to)** herstellen
**product** Produkt, Erzeugnis
**proportion** Verhältnis, Proportion
**propose (to)** vorschlagen
**public services** öffentliche Dienstleistung
**pump (to)** pumpen
**pure** rein
**purpose** Zweck

**quarter** Viertel
**quite** ziemlich

**race** Menschenrasse
**radial road** Zubringerstraße
**radio waves** Radiowellen
**railway line** Eisenbahnlinie
**rain** Regen
**rainfall** Niederschlag
**rain gauge** Regenmesser
**rainforest** Regenwald
**range** Spanne, Amplitude
**rape seed** Rapssaat
**rather** ziemlich
**raw material** Rohstoff
**ray** Strahl
**reach (to)** erreichen
**real** wirklich
**recognize (to)** erkennen
**record (to)** aufzeichnen
**rectangle** Rechteck
**reduce (to)** verringern
**reference book** Nachschlagewerk
**refinery** Raffinerie
**region** Region, Gegend
**regional airport** Regionalflughafen
**relatively** verhältnismäßig
**relax (to)** sich erholen
**relief** Oberfläche, Relief
**renewable resource** erneuerbarer Rohstoff
**republic** Republik
**residential area** Wohngebiet
**resource** Vorkommen, Rohstoff
**restaurant** Gaststätte, Restaurant
**result** Ergebnis
**right half** rechte Hälfte
**rise (to)** ansteigen
**river** Fluss
**river cruise** Flusskreuzfahrt
**road map** Straßenkarte
**root** Wurzel
**rotate (to)** sich drehen, rotieren
**rotation** Drehung, Rotation
**round** rund
**route** Route, Strecke, Weg
**run (to)** laufen, fahren, führen
**rural** ländlich

**safe** sicher
**sandy** sandig
**satellite image** Satellitenbild
**satellite map** Satellitenkarte
**satellite** Satellit
**scale (line)** Maßstab
**scenery** Szenerie, Aussicht
**sea** Meer
**seabed** Seeboden, Meeresboden
**sea birds** Seevögel
**sea level** Meeresspiegel, Meereshöhe
**sea port** Seehafen
**search (to)** untersuchen
**seaside (at the)** am Meer
**season** Jahreszeit
**seed** Samen, Saat
**service** Dienstleistung
**service centre** Dienstleistungszentrum
**service industry** Dienstleistungsindustrie
**service sector** Dienstleistungssektor
**set** hier: Paar
**shade** Schatten
**shade (to)** ausmalen
**shape** Form, Umriss
**share a border with (to)** angrenzen an
**sheep** Schaf, Schafe
**sheik** Scheich
**shopping centre** Einkaufszentrum
**show (to)** zeigen
**shower** (Regen-) Schauer
**shown** gezeigt, dargestellt
**shrub** Shrub (trockene Sträucher), Trockengebüsch
**shuffle (to)** mischen
**silly** lächerlich
**similar** ähnlich
**sink (to)** untergehen, absinken
**site** Platz, Ort
**size** Größe
**sketch map** Kartenskizze
**sketch** Zeichnung
**skills** Fertigkeiten
**skyscraper** Hochhaus
**sleet** Schneeregen, Graupeln
**slope** Hang
**Slovak** Slovake/Slovakin
**Slovenian** Slovene/Slovenin
**smile (to)** lächeln
**snow** Schnee
**soap** Seife
**so-called** so genannt
**soil** Boden
**solar system** Sonnensystem
**solution** Lösung

**solve (to)** lösen
**source** Quelle, Ursprung
**South Pole** Südpol
**Southern Hemisphere** südliche Hemisphäre
**space** Weltraum
**spaceship** Raumschiff
**Spaniard** Spanier/in
**sparse** spärlich
**speed** Geschwindigkeit
**spend (to)** verbringen
**sphere** Kugel
**spin (to)** (sich) drehen, rotieren
**split up (to)** aufteilen
**spoil (to)** verderben
**spread (to)** verteilen
**square** Quadrat
**square kilometre** Quadratkilometer
**square metre** Quadratmeter (1 m x 1 m)
**squared paper** kariertes Papier
**staff** Personal
**stage** Stufe, Rang, Level
**stamp** Briefmarke
**star** Stern
**state** Bundesland, Staat, Land
**statement** Aussage
**steep** steil
**Stevenson's screen** Wetterhäuschen
**stick (to)** kleben
**stir (to)** sich bewegen
**straight** gerade
**straw** Strohhalm, Stroh
**street map** Stadtplan
**stretch (to)** dehnen, ausdehnen, strecken
**subject** hier: Fachgebiet
**suitable** geeignet
**Sun** Sonne
**sun-bathing** Sonnenbaden
**supermarket** Supermarkt
**supply company** Zulieferbetrieb
**surface** Oberfläche, Erdoberfläche
**surface (of the sea)** Meeresoberfläche
**surround (to)** umgeben
**Swede** Schwede/Schwedin
**synthetic** künstlich

**table** Tabelle
**tanker** Tankschiff
**tape** Klebeband
**terminal (gas, oil)** Gas-, Ölanlandehafen
**temperature** Temperatur
**temperature range** Temperaturspanne
**tertiary sector** tertiärer (dritter) Sektor
**textile industry** Textilindustrie
**thematic** thematisch
**thematic map** thematische Karte
**thunderstorm** Gewitter
**tilt** Neigung, Neigungswinkel
**tilted** geneigt
**time zone** Zeitzone
**tip** Spitze
**title** Überschrift
**together** zusammen
**tool** Werkzeug
**top** Spitze, oben
**top half** obere Hälfte
**tornado** Tornado (Wirbelsturm)
**tourism** Tourismus
**tourist** Tourist
**tourist industry** Tourismus(industrie)
**tourist resort** Ferienort
**town** Stadt
**traffic** Verkehr
**transitional climate** Übergangsklima
**transparency** durchsichtige Folie
**transportation** Transport
**transportation network** Verkehrsnetz
**travel (to)** reisen
**travel agency** Reisebüro
**triangle** Dreieck
**truck** Brummi (großer Lastwagen)
**true to scale** maßstabsgerecht
**try (to)** versuchen
**tundra** Tundra
**turn into (to)** verarbeiten zu
**typical** typisch

**underground** unterirdisch
**uneven** ungleichmäßig
**Universe** Universum
**unload (to)** entladen
**upland** Hochland
**urban** städtisch
**use** Gebrauch
**usually** gewöhnlich

**vegetables** Gemüse
**vegetation** Vegetation, Pflanzenwuchs
**vertical** senkrecht, vertikal
**vessel** (See)Schiff
**view** Aussicht
**village** Dorf
**volume** Inhalt

**wall map** Wandkarte
**water** Wasser
**waterway** Wasserweg
**wave** Welle
**weather** Wetter
**wet** feucht, nass
**wheat** Weizen
**wheel** Rad
**width** Breite
**wind speed** Windgeschwindigkeit
**wind vane** Wetterfahne
**wood** Holz
**wool** Wolle
**worker** Arbeiter

**yarn** Garn
**yearly** jährlich

Bildquellenverzeichnis

Adam Opel AG, Bochum: 59 M5; ADM Oelmühle Hamburg: AG 57 M6b; agrar-press, Nörvenich: 56 M2b; Airbus Deutschland GmbH, Hamburg: 60, 60 M1, 61 M5, 61 M6, 61 M7; Alexander, Bryant u. Cherry, GB-Sturminster Newton: 34 M2; Appleby, M., Emmaus Pa, USA: 34 M3; Appleby, S., Edinburgh-UK: 62 M2, 62 M3, 62 M4, 63 M5, 63 M6, 63 M7, 64 M1, 64 M2, 64 M3, 64 M4, 64 M5, 64 M6, 64 M8; Astrofoto, Sörth: 10 M1; BASF AG, Ludwigshafen: 59 M6; Bildagentur Huber, Garmisch-Patenkirchen: 24 M2 (Lawrence), 32 M2; Canola Council of Canada: 56 M4; Centre for Bioscience, GB-Leeds: 56 M2a; Claas, Harsewinkel: 55 M4, 57 M6a; Corbis, Düsseldorf: 22 M1c (Ross), 51 M5 (Schein), 6 M1 (Wells), 7 M6 (Yamashita); Das Fotoarchiv, Essen: 34 M4 (C. Meyer); Das Luftbild-Archiv, Kasseburg: 44 M2, 45 M4; Deutsche Bahn AG: 73 M4 (Vatteroth); DLR, Oberpfaffenhofen: 38 M1; dpa, Frankfurt am Main: 35 M6, 46 M2, 52 M2; Dreymüller, B., Nohn: 8 M1; Emerson Prozess Management: 59 M8; ESA European Space Agency: 6 M7, 32 M5; Fahrzeugwerk Bronco-Höhns GmbH + Co. KG: 58 M3; Focus, Hamburg: 50 M4 (Azzi/Woodfin Camp), 6 M2 (Cosmos/Maitre); Fotoatelier Federau, Hamburg: 32 M1; Fotoflug.de, Ennepetal: 45M5; Fremdenverkehrsamt Malta, Frankfurt am Main: 68 M1, 68 M2, 68 M5; Gaffga, P., Eggenstein-Leopoldshafen: 66/67; Gartung, W., Freiburg: 33 M3; Getmapping plc, GB-Hartley Wintney: 62 M1; Greenergy International Ltd.: 57 M6d; Greiner & Meyer, Braunschweig: 33 M4 (Fiedler); Haverland, T., Soest: 35 M5; HB-Verlag, Hamburg: 6 M3; Helga Lade Fotoagentur, Frankfurt am Main: 55 M5; Hertzfeldt & Partner, Berlin: 23 M3, 26 M2, 27 M4, 28, 28 M1, 29 M4, 30 M4, 31 M7, 31 M8; IFA Bilderteam, Ottobrunn/München: 30 M1 (BCI), 31 M6 (Gottschalk), 34 M1 (Welsh); IMA, Hannover: 54 M2; Laif, Köln: 53 M6 (Reporters); Lavendelfoto, Hamburg: 56 M2c (Müller); Mauritius, Mittenwald: Titel, 46 M1 (Edmanson), 35 M8 (Fritz), 24 M4 (Hamblin), 24 M1 (Janes), 24 M3 (Plant), 46 M4 (Regener), 59 M9 (Rosenfeld); NASA, USA-Houston: 4 M1, 22 M1a, 26 M1; Paxar Central Europe GmbH, Sprockhövel: 58 M1; Peterhoff, Frank, Lenggries: 46 M3; Picture Press, Hamburg: 53 M5; picture-alliance/dpa, Frankfurt am Main: 72 M2 (Lehtikuva Oy), 72 M1 (Ossinger); picture-alliance/KPA/UTKIEK, Frankfurt am Main: 68 M6; picture-alliance/Okapia, Frankfurt am Main: 52 M4 (afp), 7 M5 (Emu), 52 M1 (Pan-Asia), ) 52 M3 (press association); picture-alliance/zb, Frankfurt am Main: 57 M6c (Förster); Reischauer, D., Hannover: 54 M1, 56 M1; Richter, U., Malsch: 30 M3; Schlemmer, H., Berlin: 22 M1b; Schönauer-Korneck, Wolfenbüttel: 45 M6a, 45 M6b, 45 M6c; Schönrock, Helmut, Bad Bentheim: 73 M5; Siemens AG, München: 58 M4; Silvestris-online, Kastl/Obb.: 31 M5; Six, R., Karlsruhe: 7 M4, Vorsatz; Stahlzentrum, Düsseldorf: 58 M2; Superbild, München: 30 M2 (Alaska-Stock); Tebbenjohanns, A., Braunschweig: 28 M2; TFA Dostmann Gmbh & Co. KG, Wertheim-Reicholzheim: 29 M5; ThyssenKrupp AG, Düsseldorf: 59 M10; ullstein bild, Berlin: 14 M1; Visum, Hamburg: 68 M3 (Vossberg); Zentrum für europäische Bildung, Bonn: 70.

# Different Types of Numbers

## 1. Reading and writing numbers

| | |
|---|---|
| . | **point** (for decimals, e.g. 2.54). See decimals. |
| , | Use a comma or a blank for groups of three numbers. (Do not read commas.) See below. |
| 0 | **zero**, "0", nought (UK) |
| 1–10 | In a text write **one** to **ten** as words. |
| >10 | In a text write numbers **larger than** 10 as numbers, e.g. "56", not "fifty-six". |
| 123 | a/one **hundred** (and) twenty-three |
| 234 | two **hundred** (and) thirty-four |
| 5,678 (5 678) | five **thousand** six hundred (and) seventy-eight |
| 93,000,000 | ninety-three **million** |
| 6,000,000,000 | six **billion** (milliard = UK-old form) |

## 2. Different types of numbers

### Ordinal numbers

| | | | |
|---|---|---|---|
| 1st | first | 7th | seventh |
| 2nd | second | 8th | eighth |
| 3rd | third | 9th | ninth |
| 4th | fourth | 10th | tenth |
| 5th | fifth | 11th | eleventh |
| 6th | sixth | 12th | twelfth |

### Fractions

| Fraction | We say ... | Decimal | Per cent |
|---|---|---|---|
| 1/2 | a/one half (of)/<br>one out of two | 0.5 | 50 % |
| 1/3 | a/one third (of)/<br>one out of three | 0.33 | 33 1/3 % |
| 1/4 | a/one fourth (of)/<br>one out of four | 0.25 | 25 % |
| 1/5 | a/one fifth (of)/<br>one out of five | 0.2 | 20 % |
| 1/10 | a/one tenth (of)/<br>one out of ten | 0.1 | 10 % |
| 1/20 | a/one twentieth (of)/<br>one out of twenty | 0.05 | 5 % |
| 1/100 | a/one hundredth (of)/<br>one out of a/one hundred | 0.01 | 1 % |

### Decimals

| | |
|---|---|
| 0.5 | (zero) **point** five |
| 3.14 (π) | three **point** one four (pi) |

### Powers/Exponents

| | | |
|---|---|---|
| $10^2$ | ten **squared**<br>ten **to the power of** two<br>ten **to the** second | 100 |
| $10^3$ | ten **to the power of** three<br>ten **to the** third | 1000 |

The Earth is **93 x $10^6$** miles from the sun.
(ninety-three **times ten to the sixth**) – 93 million –

## 3. Arithmetic

### = Proportions

| | | | |
|---|---|---|---|
| = | is, is **equal to**, **equals** | ≠ | is **unequal to**, is **not equal to** |
| > | is **more**, is **larger than** | < | is **less**, is **fewer**, is **smaller tha** |
| 4:1 | four to one | 4 x | four **times as large as** |

### + Addition → Sum

| | | | | |
|---|---|---|---|---|
| 12 | + | 4 | = | 16 |
| Twelve | **plus** | four | equals | sixteen. |

### − Subtraction → Difference

| | | | | |
|---|---|---|---|---|
| 12 | – | 4 | = | 8 |
| Twelve | **minus** | four | equals | eight. |

### X Multiplication → Product

| | | | | |
|---|---|---|---|---|
| 12 | x | 4 | = | 48 |
| Twelve | **times, multiplied by** | four | equals | forty-eigh |

### ÷ Division → Quotient

| | | | | |
|---|---|---|---|---|
| 12 | ÷ / | 4 | = | 3 |
| Twelve | **divided by** | four | equals | three. |

## 4. Geometry

| | |
|---|---|
| | **parallel lines** |
| 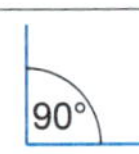  | **right angle** = 90°<br>A right angle has got ninety **degrees**. |
| 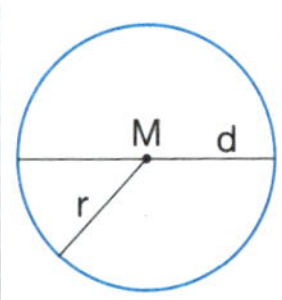  | **circle**<br>A circle is **round** or **circular**. It has got a **middle point (M)**. The line from the middle point to the circle wall is the **radius (r)**. The line from one side of the circle through the middle point to the other side is the **diameter (d)**. |
| 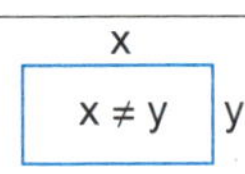  | **rectangle**<br>A rectangle is **rectangular**. It has got two pa allel equal sides on two sides at right angles |
| 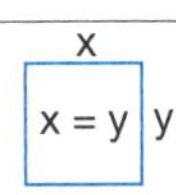  | **square**<br>A square is **square**. It has got four equal parallel sides at right angles. |
| 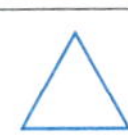 | **triangle**<br>A triangle is **triangular**. It has got three side |
| 3-d figures | cube 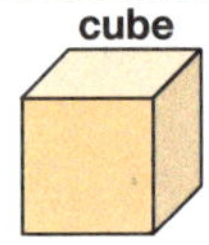 sphere 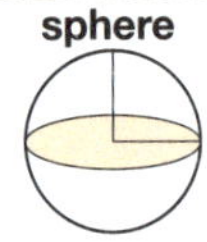 |

## 5. Time

| | | |
|---|---|---|
| **1 minute** | = 60 seconds | **2006** |
| **60 minutes** | = 1 hour | **two thousand (and) six** |
| **24 hours** | = 1 day | **twenty "o" six** |
| **365 ¼ days** | = 1 year | |
| **100 years** | = 1 century | |